Safety Assessment of Foreign Aircraft Programme

Safety Assessment of Foreign Aircraft Programme

A European Approach to Enhance Global Aviation Safety

Ronald Schnitker and Dick van het Kaar

eleven
international publishing

Published, sold and distributed by Eleven International Publishing
P.O. Box 85576
2508 CG The Hague
The Netherlands
Tel.: +31 70 33 070 33
Fax: +31 70 33 070 30
e-mail: sales@budh.nl
www.elevenpub.com

Sold and distributed in USA and Canada
International Specialized Book Services
920 NE 58th Avenue, Suite 300
Portland, OR 97213-3786, USA
Tel: 1-800-944-6190 (toll-free)
Fax: +1 503 280-8832
orders@isbs.com
www.isbs.com

Eleven International Publishing is an imprint of Boom uitgevers Den Haag.

ISBN 978-94-90947-93-4
ISBN 978-94-6094-679-0 (E-book)

Printed in The Netherlands

Table of Contents

List of Abbreviations

AAIASB	Air Accident Investigation & Aviation Safety Board
ACIP	ICAO AFI Comprehensive Implementation Programme
ACMI	Aircraft, Complete Crew, Maintenance and Insurance
AD	Airworthiness Directives
AFDD	Audit Findings and Differences Database
AFI	Comprehensive Regional Implementation Plan for Aviation Safety in Africa
AFRAA	African Airlines Association
AMC	Acceptable Means of Compliance
AMO	Approved Maintenance Organization
ANC	Air Navigation Commission
ANS	Air Navigation Services
AOC	Air Operator Certificate
ASAP	Aviation Safety Action Program
ASC	Air Safety Committee
ASIAS	Aviation Safety Information Analysis and Sharing
ATM	Air Traffic Management
BASIS	British Airways Safety Information System
BEA	Bureau d'Enquêtes et d'Analyses pour la Sécurité de l'Avion Civile
CAA	Civil Aviation Authority
CAMO	Continuing Airworthiness Management Organization
CASIS	Community Aviation Safety Improvement Strategy
CFIT	Controlled flight into terrain
CFMU	Central Flow Management Unit
Chicago Convention	Convention on International Civil Aviation
CIAA	Comité de Investigación de Accidentes Aéreos
CJAA	Central Joint Aviation Authorities
CMA	Continuous Monitoring Approach (ICAO)
COMAT	Company Materials
CRD	Comment Response Document
CRM	Crew Resource Management
CRT	Comment Response Tool

DCA	Director of Civil Aviation
DEGAS	Dutch Expert Group Aviation Safety
DGCA	Directors General of Civil Aviation
DGR	Dangerous Goods Regulations (IATA)
EASA	European Aviation Safety Agency
EC	European Community
ECAC	European Civil Aviation Conference
EEA	European Economic Area
ESARR	EUROCONTROL Safety Regulatory Requirements
ESSG	European SAFA Steering Expert Group
ESSI	European Strategic Safety Initiative
EU	European Union
EUR	European Region
EUROCONTROL	European Organization for the Safety of Air Navigation
FAA	Federal Aviation Administration of the United States
FAR	Federal Aviation Regulations
FDX	Flight Data Analysis Exchange
FOQA	Flight Operational Quality Assurance
FSF	Flight Safety Foundation
FSIX	Flight Safety Information Exchange (ICAO)
GASP	Global Aviation Safety Plan (ICAO)
GM	Guidance Material
GPWS	Ground Proximity Warning System
GSIC	Global Safety Information Center (IATA)
IAEA	International Atomic Energy Agency
IASA	International Aviation Safety Assessments (FAA)
IASDEX	International Aviation Safety Data Exchange
IATA	International Air Transport Association
IBA	International Bureau of Aviation
ICAO	International Civil Aviation Organization
ICVM	ICAO Coordinated Validation Missions
IFFAS	International Financial Facility for Aviation Safety
IOSA	IATA Operational Safety Audit
ISAGO	IATA Safety Audit for Ground Operations
ISARPs	IOSA Standards and Recommended Practices

JAA	Joint Aviation Authority
JAR-FCL	Joint Aviation Requirements for Flight Crew Licensing
MDA	Minimum Descent Altitude
MEL	Minimum Equipment List
MMEL	Master Minimum Equipment List
MOA	Maintenance Organization Approval
MTOM	Maximum Take-Off Mass
NAA	National Aviation Authority
NLR	National Aerospace Laboratory
NOTOC	Notification to Captain
NPA	Notices of Proposed Rulemaking
NTSB	National Transportation Safety Board
PANS	Procedures for Air Navigation Services
PF	Pilot Flying
PLL	Potential Loss of Life
POB	Persons on Board
QAP	Quality Assurance Program
QMS	Quality Management System
RNLAF	Royal Netherlands Air Force
SA	Safety Assurance
SAFA	Safety Assessment of Foreign Aircraft
SANE	Sustainable Aviation Network Europe
SARPs	ICAO Standards and Recommended Practices
SMS	Safety Management System
SRM	Safety Risk Management
SSP	State Safety Programme
STANAG	Standard NATO Agreements
STC	Supplemental Type Certificate
STEADES	Safety Trend Evaluation, Analysis & Data Exchange System
TCO	Third Country Operators
UN	United Nations
US	United States of America
USSR	Union of Soviet Socialist Republics
USAP	Universal Security Audit Programme
USOAP	Universal Safety Oversight Audit Programme

1 Controlling Civil Aviation Safety

1.1 History

On 17 December 1903 the Wright brothers made their historic flight at Kitty Hawk, North Carolina, USA, and were soon pursued by other aviation pioneers. Unparalleled technical advances followed these first hesitant steps of those magnificent men in their flying machines, with the result that aviation, with its international nature, has developed into a factor influencing practically every field in society.

As elementary aviation turned into commercial aviation, safety correspondingly became a considerable issue. In this context a variety of considerations have led the authorities to control aviation on military and economic grounds, but above all for safety reasons.

For many years now aviation safety has been considered to be of paramount importance. From this point of view governmental safety oversight is a significant means of ensuring application of the global Standards and Recommended Practices of ICAO.[1] In the early 1990s aviation safety assessment started to take shape, first mainly in the United States and in Western Europe, fuelled by concerns regarding the safety level of international flights to and from these regions.[2]

In mid-1991 the U.S. FAA began to formulate a programme to address these concerns, aggravated by the tragic crash of an Avianca Boeing 707 in Cove Neck, New York. The airplane ran out of fuel after repeatedly being placed in weather-related holding patterns towards the end of a flight from Bogotá, Colombia, to New York JFK international airport.[3]

---

1 According to ICAO Assembly Resolution A32-14, Appendix A, Formulation of Standards and Recommended Practices (SARPs) and Procedures for Air Navigation Services (PANS), the terms 'Standard' and 'Recommended Practice' should have the following meanings:
   a. *Standard* – any specification for physical characteristics, configuration, material, performance, personnel or procedure, the uniform application of which is recognized as necessary for the safety or regularity of international air navigation and to which contracting States will conform in accordance with the Chicago Convention; in the event of impossibility of compliance, notification to the Council is compulsory under Article 38 of the Chicago Convention; and
   b. *Recommended Practice* – any specification for physical characteristics, configuration, material, performance, personnel or procedure, the uniform application of which is recognized as desirable in the interest of safety, regularity or efficiency of international air navigation and to which contracting States will endeavour to conform in accordance with the Chicago Convention.

2 *See* IASA Process Overview of the FAA IASA Programme. *See also* COM (2002) 8 final-2002/0014(COD), submitted by the European Commission on 14 January 2002.

3 The National Transportation Safety Board concluded in its Final Report: (...) the flight crew failure to adequately manage the airplane's fuel load, and in their failure to communicate an emergency fuel situation to ATC before fuel exhaustion occurred. But the board also criticized the FAA, saying it had contributed to the accident by allowing air traffic to JFK to build up so much that day amid worsening weather that long delays were inevitable (*The New York Times*, 1 May 1991).

The preliminary safety assessment programme included visits to 12 countries, with air carriers seeking authority to operate to and from international airports located in the United States.

The initial findings convinced the FAA to establish the International Aviation Safety Assessments (IASA) programme through public policy in August of 1992, and modified it in 1994 to include public disclosure of assessment results.[4]

1.2 International Aviation Safety Assessments Program

The primary focus of this assessment programme is on a country's ability, not on an individual air carrier's, to adhere to ICAO SARPs for aircraft operations and maintenance established by the United Nations' technical agency for aviation, the International Civil Aviation Organization.

The programme, taking a top-down approach, was limited to foreign countries with air carriers providing, or wishing to engage in the operation of, international air services to and from the United States.

The purpose of the IASA programme is to ensure that all foreign air carriers of sovereign States that operate to or from the United States are properly licensed and accommodated with safety oversight, provided and performed by a competent civil aviation authority in accordance with internationally agreed ICAO standards.

Consistent with international law, certain safety requirements for operations into the United States are prescribed by the FAA Federal Aviation Regulations Part 129, Operations: Foreign air carriers and foreign operators of U.S. registered aircraft engaged in common carriage.[5] This regulation specifies that the carrier must meet the safety standards contained in ICAO Annex 6 (Operation of Aircraft) Part I (International Commercial Air Transport – Aeroplanes), 8th Edition, amended in November 2008.[6]

As a result of IASA assessments, the FAA decided in May 2000 to use only two categories or ratings for the status of countries in the future, namely:

Category 1 – in compliance with minimum international standards regarding aviation safety. A country's civil aviation authority has been assessed by FAA inspectors and has been found to license and oversee air carriers in accordance with ICAO aviation safety standards; and

---

4 *Federal Register*, Vol 57, No 164, 24 August 1992.

5 *See also* Title 14 of the Code of Federal Regulations (14 CFR) Part 129 Appendix A, FAA policy and procedures for the submission of applications for the issue or amendment of foreign air carrier operations specifications.

6 ICAO Annex 6 Part I, 8th edn, July 2001, Amendment 32, November 2008.

Category 2 – not in compliance with minimum international standards for aviation safety. A country's civil aviation authority has been found not to meet ICAO standards for aviation oversight.[7]

Countries with a Category 1 rating will be allowed normal operations to the United States, while Category 2 consists of two groups of countries: one that has air carriers with existing operations to the United States at the time of assessment. These air carriers will be permitted to continue operations at current levels under increased FAA surveillance; however, expansion or changes in services to the United States by such air carriers are not granted.

In other words, during the Category 2 rating consultation phase regarding the first group, air carrier operation into the United States will be frozen at existing levels. The second group has carriers without existing operations to the United States at the time of assessment. These carriers will not be permitted to commence services while in Category 2.[8]

Unacceptable ratings apply if:

- the civil aviation authority has not developed or implemented laws or regulations in accordance with ICAO standards;
- it lacks the technical expertise or resources to license or oversee civil aviation;
- it lacks the flight operations capability to certify, oversee and enforce air carrier operations requirements;
- it lacks the aircraft maintenance capability to certify, oversee and enforce air carrier maintenance requirements;
- it lacks appropriately trained (inspectors) personnel required by ICAO standards.

Even if carriers from a specific country are in full compliance with all international rules and regulations, they would still not be able to fly into the United States because of the lack of monitoring capabilities of its civil aviation authority.

Operations to the U.S. by a carrier from a country that has received a Category 2 rating (no permission) are not permitted unless they arrange to have their flights conducted with a duly authorized and properly supervised foreign air carrier appropriately certified from a country meeting international aviation safety standards.[9]

The FAA plans to periodically revisit civil aviation authorities of countries with air carriers operating to the U.S. to maintain full familiarity of the methods of that country's

7 Initially a third category (an unacceptable rating) was created, namely the conditional category. A country's civil aviation authority in which FAA inspectors found areas that did not meet ICAO aviation safety standards and the FAA is negotiating actively with the authority to implement corrective measures. During these negotiations, limited operations by this country's air carriers to the United States are permitted under heightened FAA operations inspections and surveillance. Since May 2000, only two categories have been applicable.

8 Available at <www.faa.gov/passengers/international_travel/>.

9 Despite the fact that Curaçao has been downgraded by the FAA to category 2, Dutch Antilles Express (DAE) will start a new service to Miami, Florida, in March 2012 by wet-leasing a MD-83 from Miami-based Falcon Air Express, which is a U.S. CAT-1 carrier (*Source*: <www.luchtvaartnieuws.nl> 26 January 2012).

continued compliance with ICAO provisions. The FAA may also find it necessary to re-assess a civil aviation authority at any time if it has reason to believe that the minimum ICAO standards are not being met.

Deficiencies found as a result of FAA assessments are almost identical to the deficiencies that have been found by ICAO during Universal Safety Oversight Audit Programme (USOAP) assessment audits. These deficiencies include:

- inadequate and in some cases nonexistent regulatory legislation;
- lack of advisory documentation;
- shortage of experienced airworthiness staff;
- lack of control on important airworthiness-related items such as issuance and enforcement of Airworthiness Directives, Minimum Equipment Lists, Investigation of Service Difficulty Reports, etc.;
- lack of adequate technical data;
- absence of Air Operator Certification (AOC) systems;
- non-conformance to the requirements of the AOC system;
- lack or shortage of adequate trained flight operations inspectors, including a lack of type ratings;
- lack of updated company manuals for use by airmen;
- inadequate proficiency check procedures; and
- inadequately trained cabin attendants.

The FAA is working towards determining that at least each country that has air carrier operations into the United States or is in the process of starting flights to the United States meets its obligations under ICAO and provides proper oversight to each air carrier operating into the United States.

The continued application of this programme will eventually result in a lower number of safety-related problems, including accidents and incidents, and an improved level of safety for the flying public.[10]

A specific initiative of the FAA that began in 1998 is the International Aviation Safety Data Exchange programme (IASDEX), an aircraft ramp inspection data exchange programme based on the ascertainment that there is no disagreement that a free flow of safety-related information among governmental regulators is indispensable to international aviation safety. IASDEX is designed to operate with any Internet connection anywhere in the world for those who have been provided secure access. This newly developed programme is expected to become an effective tool for participating civil aviation authorities, under terms of bilateral agreements between these authorities and the FAA, not only to aid them in providing oversight but also to reinforce the oversight obligations with respect to the operation of foreign aircraft within their airspace.

10 IASA Process Overview, Programs and Initiatives. About FAA.

The objective of this programme is to collect and exchange aircraft ramp inspection findings associated with foreign operators who provide services to their countries. The aim is to make full use of available safety information when performing safety oversight, including ramp inspections and surveillance of all aircraft operations, particularly foreign aircraft engaged in international air navigation.[11]

Disclosure of this kind of important safety-related information may have an impact on the safety of international air navigation and shall consequently have value to the FAA and foreign civil aviation authorities that issued Air Operator Certificates (AOC) to these operators.

Use of such special data could support participating authorities that have ICAO-compliant oversight systems in deciding whether to grant authority to foreign operators who seek to provide new air services to their countries, or allows them to take appropriate action in view of ramp inspections of foreign operators when necessary to preserve aviation safety.[12] The FAA signed the first IASDEX Memorandum of Understanding (MoU) with Canada and Australia in 2006.

1.3 UNIVERSAL SAFETY OVERSIGHT AUDIT PROGRAMME

In 1996, ICAO started to develop an improved, mandatory audit programme in order to conduct aviation safety oversight audits to identify deficiencies and encourage their resolution by States.[13] The Universal Safety Oversight Audit Programme has its origin in Resolution A29-13 (Improvement of Safety Oversight) adopted in 1992 by the 29th Session of the ICAO Assembly. This Assembly Resolution noted the concern that some contracting States might experience difficulties in accomplishing their safety oversight obligations, one of the principles of the Chicago Convention.

The Air Navigation Commission (ANC) recommended that assessments be focused solely on a State's oversight capability based on responsibilities in Annex 1 Personnel Licensing, Annex 6 Aircraft Operations, and Annex 8 Airworthiness, and should be mandatory. In addition, it called for transparency of the assessment results. However, many Commission members were of the opinion that contracting States were more likely to accept the programme if it were voluntary and confidential, like the earlier issued ICAO

11 ICAO NACC/DCA/2-WP/07, 18 August 2005, Second meeting of North American, Central American and Caribbean Directors of Civil Aviation, Tegucigalpa, Honduras, 11-14 October 2005, International Aviation Safety Exchange (IASDEX) Programme.

12 *Id.*

13 Safety oversight is a function by which States ensure the effective implementation of the safety-related Standards and Recommended Practices and associated procedures contained in the Annexes to the Convention on International Civil Aviation and related documents (ICAO definition in *Safety Oversight Manual, Part A, The Establishment and Management of a State's Oversight System*, paragraph 2.1.1, ICAO Doc 9734, AN/959, 2nd edn, 2006). *See also* another explanation of safety oversight in note 1 (Chapter 6).

assessment programme regarding aviation security. Resolution A29-13 called on States to provide financial and technical resources to enable other States to perform their safety oversight responsibilities regarding air carrier operations. After intense debate it was finally agreed that the final assessment reports would be confidential, but that upon request a summary report noting deficiencies in implementation of the relevant standards would be available to contracting States in question.[14]

On 7 June 1995, the ICAO Council approved this original programme (Safety Oversight Program) as well as the related mechanism for financial and technical contributions. The realization was generated in part as a response to the IASA programme of the FAA. In addition, the United States Government supported the programme's development.[15]

This Safety Oversight Program, originally based on a Dutch initiative, among other things, was endorsed by the 31st Session of the Assembly and became operational in March 1996. It was to be considered the precursor of the subsequent USOAP.[16] The Safety Oversight Program included not only assessments, but also a training programme for team members, safety oversight seminars and workshops conducted by ICAO for the benefit of contracting States and the follow-up activity, which frequently included the ICAO Technical Cooperation Bureau.

The original ICAO Safety Oversight Program had four primary characteristics:

1. The assessment;
2. The programme was funded by State contributions;
3. The programme was voluntary;
4. The findings were confidential.

The year 1996 was the worst in civil aviation history in terms of fatalities in aircraft accidents. In particular, the entire African continent suffered from perceived unsafe conditions in the air navigation system.

This situation, along with increased concern over safety worldwide and various assessment reports on lack of implementation and non-compliance with safety requirements, made ICAO focus on a series of safety questions and provided the atmosphere to concede the development of a more aggressive safety oversight programme.

In 1996 ICAO began to develop an enhanced programme to introduce regular, mandatory safety audits on all contracting States, to be carried out by ICAO, to provide for greater

14 J.H. Loos, The ICAO Safety Oversight Program, The George Washington Aviation Institute, 1999.

15 The words 'programme' and 'program' are used alternatively, depending on the original denomination and use.

16 ICAO Oversight Programme. On 7 June 1995 the ICAO Council approved this original programme as well as the related mechanism for financial and technical contributions. The programme was endorsed by the 31st Session of the ICAO Assembly and became operational in March 1996. *See also* ICAO Assembly Resolution A29-13 Improvement of Safety Oversight, adopted in 1992 by the 29th Session of the ICAO Assembly. The title 'Safety Oversight Program' (SOP) is a generic term covering both the original Safety Oversight Program and the new, improved version, called ICAO Universal Safety Oversight Programme.

transparency in the dissemination of audit results, to ensure the allocation of adequate ICAO funds and to expand the programme to other technical fields at the appropriate time.

In 1997, Dr. Assad Kotaite, President of the ICAO Council, gave a speech in Washington, D.C., USA, about the possibility of a more aggressive approach by ICAO to safety assessment audits.[17] After various discussions, the Directors General of Civil Aviation convened in Montreal on 10 November 1997, the first time that ICAO had ever called such a meeting (Conference on a Global Strategy for Safety Oversight), and reaffirmed the need for safety oversight. In fact, they formulated a global strategy for improved safety oversight based on practical and concrete steps for implementation by each contracting State of an effective safety oversight system, consistent with the ICAO provisions. The Directors General of Civil Aviation reached an agreement on 12 November to initiate a universally applicable audit programme (USOAP), which would have the following specific characteristics:

1. It would be an audit rather than an assessment;
2. It would be universal and mandatory;
3. It would have a much larger degree of transparency; and
4. It would be funded from the ICAO budget.

The Council approved the improved programme on 6 May 1998, and the 32nd Session of the Assembly in Montreal (September–October 1998) endorsed the proposal in Resolution A32-11 (Establishment of an ICAO universal safety oversight audit programme) and A32-24 (Budgets 1999, 2000 and 2001).

The improved audit programme was launched in January 1999 as a transparent audit programme comprising regular, mandatory, systematic and harmonized safety audits.

The methodology of the Universal Safety Oversight Audit Programme, as it was called, consists of a top-down approach, concentrating on systematic auditing and monitoring of the ability and competence of individual States to conform to the safety-related ICAO SARPs.[18]

ICAO prescribes eight critical elements of safety oversight in its own guidance material (Safety Oversight Manual, Part A: The Establishment and Management of a State's Safety Oversight System – Doc 9734). These critical elements are:

1. Primary aviation legislation
2. Specific operating regulations
3. Civil Aviation Authority (CAA) structure and safety oversight functions
4. Technical guidance
5. Qualified technical personnel

---

17 Speech during the White House Commission International Conference on Aviation Safety and Security, 13-15 January 1997, George Washington University.

18 An ICAO USOAP audit is much more thorough than an assessment. Where the method of assessment concentrates on ICAO Standards and Recommended Practices, the more comprehensive audit also considers existing guidance material and industry practice.

6. Licensing and certification obligations
7. Continued surveillance obligations
8. Resolution of safety issues

This Safety Oversight Manual is directed at 'high-level government decision makers' as it highlights the obligations of contracting States under the Chicago Convention and offers guidance on how to establish a compliant oversight system. Given the circumstances, the ICAO audit programme should be aligned with this document.[19]

'Universal and mandatory' means that it will apply to all contracting States and will make it mandatory with the consent of the State to be examined by means of a Memorandum of Understanding, which might seem a contradiction in terms, but obviously with the intent to meet the principle of sovereignty.

Before introducing the less formal concept of a MoU in this context, the option to circumvent onerous legal obligations under international law, ICAO was initially examining three possibilities for obtaining the consent of the States for introducing a system of mandatory audits. Such consent might have been achieved through:

- A bilateral instrument;
- A multilateral instrument; or
- A combination of these two instruments, such as an Assembly resolution approving the ICAO audit programme, followed by individual bilateral instruments between States and ICAO.

> Whereas the 32nd Assembly urged all contracting States to agree to audits to be carried out upon ICAO's initiative, but always with the consent of the State to be audited, by signing a bilateral Memorandum of Understanding with the Organization.[20]

The USOAP audits include industry visits; however, the focus is on governments, especially the civil aviation authorities of the contracting States concerned. The audits consist of:

1. Compliance with the Chicago Convention and State Regulations;
2. Conformance with ICAO standards;
3. Adherence to recommended practices, related procedures, guidance material and relevant industry practices in general use.

---

19 ICAO Assembly Working Paper A33-WP/66 EX/23 Enhancing the Quality and Usefulness of ICAO Audit Information, 25 July 2001.

20 ICAO Assembly Resolution A32-11, dated 1 October 1998, forms the basis of the ICAO policy and objectives for conducting safety oversight audits. As such, the objective of the USOAP is to promote global aviation safety by conducting safety oversight audits of all contracting States. Effective implementation of the USOAP required that fundamental auditing principles be developed and established to guide ICAO's auditing activities and to assure States of the validity of the audit process.

The ICAO audit teams are looking specifically at:

1. National aviation legislation and specific regulations;
2. Organizational structure and legal status of the State's CAA;
3. System for certification and continued surveillance of aircraft, personnel and operators; and
4. Implementation of SARPs.

The safety oversight audit team will review the State's regulatory provisions, examine records and documentation and conduct interviews, in order to make its findings and recommendations, on the basis of safety oversight audit general requirements, as follows:

a. consideration of existing national legislation enabling States to enforce the provisions of the Chicago Convention;
b. consideration of existing national regulations enabling States to carry out detailed safety regulation in support of the Chicago Convention and Annexes 1, 6 and 8 thereto and related provisions in other Annexes, as well as associated ICAO Procedures for Air Navigation Services (PANS) and other procedures, guidance material and relevant safety-related practices in general use in the aviation industry as referred to in such material;
c. a system for both the certification and the continued surveillance of personnel, operations of aircraft, air navigation services, aerodromes and ground aids, the safe transport of dangerous goods by air, as well as a system for accident and incident investigation, in particular with respect to:
    1. qualifying aircraft operations aircraft maintenance and air traffic services personnel, aerodrome personnel, and accident and incident investigators supported by an appropriate organizational and managerial structure;
    2. detailed records of training programmes and actual training given to professional staff in the safety-related area of the industry;
    3. availability of appropriate reference material, including ICAO documentation;
    4. information on which ICAO SARPs and other regulations, as applicable, have been used for the licensing of personnel and the certification of approved maintenance organizations, aerodromes, etc.;
    5. safety management systems and safety management programmes in place;
    6. inspection or surveillance results containing details of technical and operational deficiencies;
    7. evidence of procedures for the issuance and approval of licences and certificates and their suspension and revocation when unsafe conditions are identified, and records of how often action has been taken; and
    8. evidence of overall system safety awareness and of a mechanism for accident and incident prevention; and

d. implementation of ICAO SARPs, PANS and other procedures, as well as adherence to guidance material and relevant safety-related practices in general use in the aviation industry as referred to in such material, related to the operation of aircraft, airworthiness of aircraft, air navigation services, aerodromes and ground aids, accident and incident prevention and investigation, the safe transport of dangerous goods by air, and the safety-related activities required by all safety-related Annexes, in particular with respect to the requirements of an adequate organization, training programmes, method of control and supervision of personnel licensing, aircraft operations, airworthiness of aircraft, air traffic and aerodrome operations, accident investigations and the safe transport of dangerous goods by air consistent with the nature and extent of the operations specified.[21]

ICAO audits have revealed organization-related problems, arising mainly from a lack of commitment by governments to support their corresponding civil aviation authorities at the desired level. These problems are manifested as inability to attract, recruit and retain adequately qualified and experienced personnel, resulting in high technical staff turnover; job insecurity; low staff morale and the often active search for additional employment; and the lack of adequate support staff, equipment, facilities and guidance material. Although States in general have submitted an acceptable corrective action plan to resolve the problems identified, the fact remains that several States may experience difficulties in the implementation of their corrective action plans for various reasons, including the lack of adequate financial and qualified human resources.

Inadequate funding to civil aviation authorities has long been identified as a reason for the deficiencies identified by ICAO audits and the difficulties experienced by civil aviation administrations. ICAO audits indicate that most civil aviation authorities are capable of generating adequate funds to enable them to effectively undertake their responsibilities. However, in many cases, funds raised are not directly reinvested in civil aviation activities, thus depriving civil aviation authorities of a viable source of funding for conducting their safety-related responsibilities in the most effective and efficient manner.

The audits have revealed that in a number of States, the budget allocated to civil aviation authorities represented <10% of the funds generated through civil aviation activities. However, a number of the problems can be resolved with the assistance of ICAO through the provision of support at a global or regional level. This support would include follow-up missions; the development and delivery of training courses, seminars and workshops; and the development of safety oversight-related guidance material, including relevant handbooks and checklists.[22]

---

21 ICAO Doc 9735 AN/960, *Safety Oversight Manual*, 2nd edn, 2006, Appendix B Generic Memorandum of Understanding, paragraph 12.

22 ICAO Assembly Working Paper A33-WP/49 EX/7 Assisting States to Resolve Deficiencies Identified by the Safety Oversight Audits, 9 July 2001, at 2, 9.

Subjected to ICAO audits, contracting States indicated that they had benefited from the audits in several ways. Besides helping them to identify shortcomings within their safety oversight systems, the audits have also served to raise the awareness of high-level decision-makers to provide the civil aviation authorities with the required funds and support.

Audited contracting States have been highly supportive of USOAP and have shown increasing interest in the programme, as manifested through their acceptance, with no exception, of the audit missions on the basis of the MoU approved by the Council and their willingness to submit a corrective action plan within the specified period. As a matter of fact, they have been unanimous in their recognition of the ICAO USOAP as an essential tool that should continue to exist as one of the major activities of the Organization.

USOAP has shown that accurate, factual and timely information enables all contracting States and the aviation community, in general, to develop a strategy to assist in the resolution of identified safety concerns.

1.4 IATA OPERATIONAL SAFETY AUDIT PROGRAMME

This ICAO USOAP audit programme and FAA's IASA assessment (consultation) programme are focusing on State conformance with ICAO Annexes, while the IATA Operational Safety Audit (IOSA) programme is primarily designed for auditing international air carriers to identify latent failures.[23]

IOSA was established in 2001 to meet air carrier industry needs for common globally harmonized operating safety standards. IOSA, like the other assessment and audit programmes, is also based on ICAO standards, and is indirectly providing a measure of State oversight compliance via air carrier operating performance assessments. IOSA is not intended as a substitute for State regulatory authority oversight. The specific aviation industry needs in the areas of cost-effectiveness and safety are as follows:

- The air carrier industry is subject to an ever increasing proliferation of audits that often overlap both in intent and content. IOSA will change that by introducing a single, agreed, common air carrier audit standard.
- Safety of air carrier operations remains the highest industry priority. Compliance with IOSA Standards and Recommended Practices (ISARPs) will help an air carrier to deliver operational safety in all key areas, as well as operating more efficiently.

IOSA provides a standardized audit programme based on internationally recognized standards and a structured system for the sharing of audits. The IOSA standards, on which the audits are based, have been developed by a diverse and highly experienced group of air carrier audit specialists, under the guidance and direction of IATA. IATA member air

23 IATA created IOSA in 2001, focusing on air carrier organizations, operational management and control systems.

carriers are committed to being audited to IOSA standards. A growing number of regulatory authorities worldwide are beginning to see how IOSA can help them by extending and enhancing their regulatory oversight role. Non-IATA member air carriers are starting to see the benefits that can accrue from aligning their operations with IOSA.[24]

As innovative technology and human endeavours have already contributed significantly to the aviation safety record, the most important contemporary safety strategy concentrates on the relationship between management and safety. IOSA as an evaluation system is designed to assess the operational management and control systems of an air carrier. IATA has the opportunity to outsource the conduct of safety audits to IATA Accredited Audit Organizations (AO). As a matter of fact, IOSA has become a more comprehensive and complementary programme to the other audit programmes and has become integral to the safety management systems of a number of international air carriers, but above all has become the agenda setter for safety specialists within those air carriers while earning endorsements from civil aviation authorities.[25] Registration is the final objective of the IOSA audit process. An air carrier that has demonstrated full conformity with the IOSA standards will be entered in the IATA registry as an IOSA operator. Audit reports are only accessible following authorization by the air carrier and application grant of IATA.

With the implementation and international acceptance of IOSA, air carriers and regulators achieve the following benefits:

- A reduction of costs and audit resource requirements for air carriers and regulators.
- Continuous updating of standards to reflect regulatory revisions and the evolution of best practices within the industry.
- A quality audit programme under the continuing stewardship of IATA.
- Accredited audit organizations with formally trained and qualified auditors.
- Accredited training organizations with structured auditors training courses.
- A structured audit methodology, including standardized checklists.
- Elimination of audit redundancy through mutual acceptance of audit reports.
- Development of auditor training courses for the air carrier industry.

Apart from any organizational safety programme, whether it is IOSA or SAFA or any other safety deficiency verifying system assessing or auditing safety aspects of air carriers or governmental civil aviation authorities, a basic action programme of principles is required. Therefore the following should be addressed:

- Conduct an audit in such a way as not to be seen to have negative implications. It must be recognized as providing positive and valuable information as the basis of management action towards measurable improvements in safety;

---

24 <www.iata.org/iosa>.

25 W. Rosenkrans, *Delving into IOSA*, AeroSafety World, Flight Safety Foundation, December 2010-January 2011, at 13. Available at <http://flightsafety.org/aerosafety-world-magazine/december-2010-january-2011/delving-into-IOSA>.

- Cover all divisions or bases of a company's operation and their interlocking aspects. Naturally, air safety audits are varied in scope and format depending on the individual type of organization or company;
- Monitor ongoing operations in the context of adherence to policies, procedures and standards;
- Interview the appropriate organization or company representatives to become aware of the company's policies, organizational structure, operations, personnel, equipment, reporting structures, facilities, regulatory constraints, training systems and any other relevant information;
- Identify any specific details of the organization or company's operations that will require special intention;
- It is essential that full agreement and understanding is reached by the organization or company management being audited concerning systematic mandatory auditing, on the scope and terms of references of the audit including any specific exclusions, and subsequently to make the organization or company aware of any limitations this imposes on the effectiveness of the audit;
- Review of all relevant manuals and regulations. Pre-audit any copies obtained as references throughout the audit;
- Develop and present an audit program and deadlines, and detail the input required from the organization or company;
- Management must publicize and support it in information to staff members;
- Personal and confidential interviews are required with employees at all levels of the organization or company. This is a delicate procedure and requires sensitivity and an ability to build trust and confidence. Staff must be encouraged to speak honestly with the audit team and must be convinced this will not be considered disloyal. Interviewers should note areas that interviews suggest require examination in depth, and those interviewed must be assured their input will be confidential and that the source will remain unidentifiable.

The categories under which problems and solutions are likely to be identified and tend to interact are generally organizational characteristics, policies and procedures, communication, morale and discipline, and safety management and organization of safety activities.

Given the massive growth of air transportation worldwide, with the associated potential for increased risk of accidents, flight safety inspection methods used so far, whether they comprise bottom-up methods such as IOSA and SAFA, or top-down methods such as USOAP and IASA, have yielded impressive results and gains in aviation safety. They have drastically reduced world accident ratios and the number of fatalities.

1.5 International Civil Aviation Organization

Recently, ICAO has been working on principles to prevent the improper use of information obtained by countries from the ramp inspections on foreign aircraft.[26]

A potential risk exists to misuse information gained from inspections of foreign aircraft in dishonest competition between operators.

The results of these inspections are used by some countries to impose sanctions on the activity of operators from other countries. Therefore, coordination and the adoption of generally accepted procedures and conditions for applying sanctions to operators are required to improve the effectiveness of safety assessment programmes to supervise the activity of operators.[27]

A general but worrying overview of the problems identified by safety oversight audits includes:

- lack of an appropriate legislative framework, including lack of basic civil aviation law, appropriate regulations, procedures, documentation and guidance material;
- lack of an appropriately established and funded civil aviation organization and safety management system;
- lack of appropriate qualified and experienced technical personnel; inadequate certification and licensing systems;
- lack of a system for the control and supervision of licensed and certificated bodies; and
- lack of a basic surveillance system; and lack of a system for the resolution of safety issues.

In the interest of improving civil aviation safety throughout the world, ICAO will undoubtedly be the designated global organization to create a vehicle for the integration of global oversight audit and ramp inspection information. Such critical safety information can be utilized to enhance individual proactive oversight audit and ramp inspection programmes such as USOAP, IASA as well as IASDEX, IOSA and SAFA, and integrate the resulting safety information into the ICAO Continuous Monitoring Approach (CMA) as well as ICAO Coordinated Validation Mission (ICVM) programmes, with the clear objective of facilitating the protection, analysis and dissemination of this safety data, to prevent unnecessary duplication and to enable the civil aviation industry to reduce worldwide aircraft accident rates.[28]

---

26 ICAO Working Paper High-Level Safety Conference, HLSC 2010-WP/78, Montreal 29 March-1 April 2010, Doc 9935.

27 ICAO Assembly Working Paper A37-WP/273 TE/146, 27 September 2010, Assembly-37th Session, Technical Commission. Coordinating the Activity of Participating Countries when Implementing Safety Assessment Programmes for Foreign Aircraft and Using the Results of these Inspections, at 1.

28 ICAO Working Paper High-Level Safety Conference, HLSC 2010-WP/78, Montreal 29 March-1 April 2010, Doc 9935.

ICAO has developed an Audit Findings and Differences Database (AFDD), based on factual and accurate data, to enable it to analyze the audit findings and thus accurately identify deficiencies impacting on safety, consistent with regional accident and incident rates, and also to determine actions required to resolve the safety concerns. It functions as a reliable safety tool, both for ICAO and its contracting States. The analysis enables the identification and quantification of safety concerns at individual State, groups of States, regional and global levels.

In a practical overview, the data collected enables ICAO to:

a. have factual and accurate data on the level of aviation activity: globally, by region, by State or by a group of States;
b. determine the types of difficulties experienced in each of the five protocol areas: Legislation and Regulations, Organization of the State's Civil Aviation Administration, Personnel Licensing and Training, Operations of Aircraft and Airworthiness of Aircraft, based on the type of activity or status of implementation, as applicable, for an individual State or group of States at a regional and global level;
c. identify major safety concerns on standards and procedures for issuing air operator certificates, personnel licensing and certificates of airworthiness;
d. determine the percentage of findings *vis-à-vis* each provision contained in the Annexes, or guidance material: globally, by region, by State, or by a group of States;
e. identify major safety concerns and determine their impact on the safety of aircraft operations at various levels, with the objective of enabling ICAO, international and regional organizations, groups of States and/or individual States to prioritize actions directed at resolving identified and quantified safety concerns;
f. determine the level of implementation of Annex SARPs in the audited areas, classified as follows: exceeds ICAO SARPs; other means of compliance; not-implemented and not-applicable;
g. compare the types of findings generated with the audited differences identified, to determine, on the basis of factual data, the validity and accuracy of an existing SARP, or the need to develop further guidance material; and
h. have factual and accurate information for the use of ICAO Regional Offices to enable them to prioritize their activities in the audited areas.[29]

In the meantime the programme has shown that accurate, factual and timely information enables all contracting States and the aviation community, in general, to develop a strategy to assist in the resolution of identified safety concerns.

The experience built up in the course of the audits and the significant volume of information that came from States unequivocally show that the ICAO USOAP is one of the

29 ICAO Assembly Working Paper A33-WP/47 EX/10, 9 July 2001. *See also* 18th E/CAR DGA-IP/03, 17 November 2003.

most important programmes ever developed by the Organization to enhance flight safety. However, the experience of the long-term funding of the programme and its integration into the Organization's regular work programme, as well as the conclusions drawn with respect to the findings of the safety oversight audits, show the need for a certain review of the scale and volume of the future activities to be undertaken within the framework of the programme.

In order to further enhance flight safety at the national, regional and global levels, it appears advisable in the long-term programme of audits to provide for the transition from formal audits of factors influencing flight safety to direct audits of compliance by States with a certain assigned target level of flight safety, recommended by ICAO for global compliance.[30]

In recognition of the success of the USOAP, during the 35th ICAO Assembly, a proposal was handed in for the continuation and expansion of this progressive programme as of 2005 and it was agreed, with Assembly Resolution A35-6, that the extension should include all safety-related provisions in all safety-related Annexes, except for Annex 9 – Facilitation and Annex 17 – Security.[31]

The USOAP has proven to be a comprehensive safety evaluation tool. In a period of about five years, ICAO has changed, by conducting a large number of safety audits under this programme, to an organization aggressively pursuing the implementation of its SARPs and procedures. Evaluation will be extended at the appropriate time to include airports, air traffic services and support facilities and services.[32]

1.6 Safety Assessment of Foreign Aircraft Programme

In 1996, in the wake of the Puerto Plata fatal aircraft accident (Birgenair Flight 301), a Communication from the European Commission to the European Council and Parliament devised and defined a Community Aviation Safety Improvement Strategy (CASIS) by convening a high-level group of aviation safety experts who assisted in elaborating this plan.[33] This safety improvement strategy was based on two main streams of actions: concerning safety of operations within Europe, the creation of EASA, and concerning actions *vis-à-vis* third countries, drafting a SAFA Directive and at the same time developing

30 ICAO Assembly Working Paper A33-WP/112 EX/39 Need for the Development of the Safety Oversight Concept and a Long-Term Work Programme in This Field, 29 August 2001.

31 *See also* ICAO Working Paper NACC/DCA/2-WP/04, 31 August 2005, Safety Oversight Matters: Progress Report on the Implementation of the ICAO USOAP, paragraph 4. Transition of USOAP to a Comprehensive Systems Approach, at 2. *See also* ICAO Assembly Resolutions A33-8 and A33-9 (2001).

32 ICAO Assembly 35th Session (Montreal, 28 September-8 October 2004). *See also* ICAO Assembly Working Paper A36-WP/64 EX/22 Progress Report on the Implementation of the ICAO USOAP under the Comprehensive Systems Approach, 2 August 2007. Development of USOAP beyond 2010 at 5.

33 Doc SEC (96) 1083-C4-0423/96.

technical assistance for States to fulfil their ICAO obligations with respect to safety standards, if necessary. The Community Aviation Safety Improvement Strategy was eventually adopted and transmitted to the Council on 12 June 1996.[34]

The overall objective was to accomplish a comprehensive contribution to global aviation safety enhancement within the framework of the Chicago Convention and use the available and applicable Community instruments.

In the same year, ECAC launched a voluntary safety assessment programme, called ECAC SAFA, which was complementary to but slightly different from the ICAO audits.[35] This aviation safety assessment programme was based on the understanding that maintaining confidence in the safety oversight provided by other States was a prerequisite for the continued development of a well-functioning and reliable global air transport system. Like IOSA, ECAC SAFA is based on a bottom-up approach, basically taking as its starting point random ramp inspections of commercial aircraft landing at international airports located in ECAC States, in order to be able to oblige operators and appropriate civil aviation authorities to take corrective measures in specific safety cases showing significant deficiencies and irregularities.

A logical consequence of these ramp inspections and subsequent corrective actions in a bottom-up approach procedure was the involvement of States of Registry and States of the Operator, as dictated by circumstances.

The main features of the ECAC SAFA programme are:

- Its application by all 44 member States including the sharing of information between the regulatory authorities, through a centralized database;
- Its bottom-up approach: the core of the programme is built around ramp inspections of aircraft;
- Its non-discriminatory nature. It applies to aircraft from ECAC and non-ECAC States without any form of inequality.

Following a decision of the Directors General of ECAC member States, the former ECAC SAFA programme was transferred under the competency of the European Commission, assisted by EASA, as of 1 January 2007. A major feature of the transition process was to maintain and intensify the pan-European spirit of SAFA, which was finalized by the conclusion and signature of a working agreement between EASA, being a European Community institution, and the non-EU ECAC member States, who previously joined the voluntary ECAC SAFA programme and would continue to participate in the EC SAFA programme.

Contrary to the new set-up of the EC SAFA programme, the former voluntary ECAC SAFA programme was based not upon a European legal binding basis but upon a commitment of the Directors General of Civil Aviation of the participating ECAC member

34 Doc SEC (96) 2083 final.

35 ICAO Annexes 1, 6 and 8 served as references for the ECAC SAFA foreign aircraft ramp inspections.

States. However, as per 30 April 2004, Directive 2004/36/EC created in this respect a legal obligation upon EU member States to perform ramp inspections on third-country aircraft landing at Community international airports (including military airfields in the European region (EUR), open to international air traffic and provided with facilities to dispatch commercial aircraft).[36]

One of the considerations in establishing this Directive was the Resolution regarding the air disaster off the coast of the Dominican Republic, which was adopted by the European Parliament on 15 February 1996, endorsed by the Council Conclusion of 11 March 1996, indicating the need for the Community to take a more active stance and develop a strategy to improve the safety of its citizens travelling by air or living near airports.[37]

Stakeholders participating in the EC SAFA programme are: the European Commission, EASA, the member States and EUROCONTROL. They regularly join meetings of the Air Safety Committee (ASC) and the European SAFA Steering Expert Group (ESSG). In the ESSG, the non-EU ECAC member States engaged in the EC SAFA programme together with the EU member States are represented having a technical advisory role regarding, in particular, through performance of the following tasks:

- Assisting the European Commission (in a consultative role) in relation to proposals from EASA prior to their submission to the ASC, regarding the overall safety oversight policy of the EC SAFA programme, possible actions to be taken in specific safety cases and recommendations for the further development of the EC SAFA programme, including the amendment of SAFA procedures;
- Communicating to the Commission feedback upon the national implementation of the EC SAFA programme in their State;
- Reviewing and reacting upon the SAFA database analysis that EASA is required to conduct on a frequent basis.[38]

---

36 *See* DGRLD/LI/DIR/00.190060, 18 April 2000, at 1.

37 Resolution B4-0150/96, *Official Journal of the European Communities* C65, 4 March 1996, at 172 (*See also* COM (97) 55 final, 17 February 1997). The aviation accident, which this Resolution is referring to is Birgenair Flight 301 from Puerto Plata via Gander to Berlin and continuing to Frankfurt, chartered by Turkish-managed Birgenair partner Alas Nacionales. On 6 February 1996 the Boeing 757-225, registration mark TC-GEN, operating the route to Europe, crashed shortly after take-off from Puerto Plata's *Gregorio Luperón* international airport located on the north coast of the Dominican Republic. The probable cause of the accident, according to the Dominican Republic government's *Dirección General de Aeronáutica Civil* (DGAC), was: The crew's failure to recognize the activation of the stick shaker as a warning of an imminent entry into a stall, and the failure of the crew to execute the procedures for recovery from the onset of loss of control. The investigation concluded that one of the three pitot tubes, used to measure airspeed, was blocked. The aircraft had been parked at the airport and not flown for 25 days, during which time the pitot tubes were not covered and possibly vulnerable to creating clogging insect nests inside the pitot tubes.

38 Article 2, Commission Regulation (EC) No 768/2006 of 19 May 2006 implementing Directive 2004/36/EC of the European Parliament and of the Council as regards the collection and exchange of information on the safety of aircraft using Community airports and the management of the information system, *Official Journal of the European Union L* 134/16, 20 May 2006.

Any infrastructure, product, system, equipment, service, person or organization not covered by Community competence will remain under the full responsibility of the member States, which have to take, as appropriate, measures to provide for the level of protection expected by their citizens.

The European Commission, as the executive body of the European Union,[39] carries overall responsibility and has legislative powers, while EASA is responsible for some specific executive tasks in the field of a centralized database as well as analyzing, advising and reporting of safety information and follow-up actions, strategy and developing procedures as well as training programmes.

A more detailed overview of the specified role and responsibilities of EASA in the EC SAFA programme includes:

- Collect by means of a centralized database the inspection reports of the member States engaged in the EC SAFA programme;
- Maintain and further develop the centralized database;
- Analyze all relevant information concerning the safety of aircraft and its operators;
- Report potential aviation safety problems to the European Commission and the member States;
- Advise the European Commission and the member States on follow-up actions;
- Advise on the future development and strategy of the EC SAFA programme;
- Develop SAFA procedures;
- Develop training programmes and foster its organization and implementation.[40]

The SAFA centralized database has been managed by EASA since December 2006, when it was transferred from the Central Joint Aviation Authorities (CJAA) in Hoofddorp, the Netherlands, to EASA in Cologne, Germany.[41]

The conduct, reporting and follow-up of SAFA ramp inspections are carried out according to common procedures established at the Community level by Directive 2004/36/EC. From the beginning, the Community fully endorsed and supported the SAFA programme by means of participating through its membership of the SAFA ESSG and by providing finance to the Joint Aviation Authorities.

In order to accomplish and maintain a high uniform safety level in this specific context, a legal basis was established in 2004 by adopting this SAFA Directive, in 2005 complemented by Regulation (EC) No 2111/2005 on the establishment of a list of air carriers subject to an operating ban within the European Community airspace.

---

39 The European Commission occupies a unique administrative position in the Community: it is a body that proposes community legislation; it is the executive body and the supervisory body as well as the representative body.

40 EASA Approvals and Standardization Directorate.

41 COM (2009) 627 final, at 2.

> An operating ban means the refusal, suspension, revocation or restriction of an air carrier's operating authorization or technical permission for safety reasons, or any equivalent safety measures in respect of an air carrier which has no traffic rights in the Community but whose aircraft might otherwise be operated in the Community under a lease agreement.[42]

To best achieve the objectives of the EC SAFA programme, including unremitting compliance with ICAO safety standards and adequate and proper governmental safety oversight, close cooperation with the civil aviation authorities of all those States whose operators and aircraft have been subject to SAFA inspections is imperative.

EASA, as an agency that has been given specific regulatory and executive tasks in the field of aviation safety, is playing a major role regarding the EC SAFA programme.

1.7 SAFA, EASA and ICAO

Aviation safety can be effectively improved only if relevant organizations have access to sufficient resources. ICAO's International Financial Facility for Aviation Safety, to be financed from contributions from contracting States, is intended to be used to improve aviation safety through recovery actions, recommended as a result of ICAO's audits carried out under the USOAP. Major beneficiaries would be the less developed nations of the world.

Since the European Union is having its own priorities for recovery actions, the Commission considers that the Community and its member States should play a much more proactive role with respect to specific regions and will use part of the significant resources they allocate to technical assistance to developing countries to support aviation safety programmes with the aim of enhancing aviation safety.[43] The Community has a number of financial instruments and a list of currently supported safety actions.[44]

Drawing lessons from the new role of ICAO in the safety oversight auditing of its contracting States, the European Commission is envisaging a new approach to ensure that funds will be used effectively to provide sustained improvement in aviation safety, and will simultaneously facilitate complementarities of various actions with the same aim undertaken by all other interested parties:

- Any cooperation activity shall normally be the follow-up of an ICAO assessment conducted in the framework of the USOAP, unless an emergency action is identified as

42 Article 2, Definitions, (g) of Regulation (EC) No 2111/2005.

43 Safety Programme: An integrated set of regulations and activities aimed at improving safety (ICAO Annex 1 Chapter 1 1.1 Definitions).

44 Community funds available for World Aviation Safety Improvement: Economic Development Fund (Lome countries), Overseas Development and Economic Co-operation Budgets, TACIS (Technical Aid to the Commonwealth of Independent States), PHARE (originally created as the Poland and Hungary Assistance for Restructuring their Economies plan) and the Budget for Potential New EU member States.

necessary before the ICAO assessment takes place, for which some flexibility of approach will need to be maintained;
- Any project shall be based upon the ICAO Approval Action Plan established as a result of the audit;
- Each project shall be discussed and agreed between the European Commission, the recipient country and ICAO to ensure that it provides an efficient answer to the difficulty identified by ICAO;
- European industry will be consulted on the projects envisaged and invited to contribute if they so wish;
- Recipient countries shall contribute to the implementation of the projects and commit to taking the necessary measures to ensure their sustainable continuation;
- The terms of reference of the project shall be established with appropriate technical support from aviation experts;
- The implementation of the project will have to be carried out in accordance with the rules of procedure applicable to the eligible fund; when a call for tender is required, the selection should be done using the necessary aviation expertise;
- The implementation also needs continuous monitoring by appropriate aviation experts;
- At the end of the implementation, there shall be another ICAO audit in order to check that the result meets the expectation and that the relevant part of the action plan is achieved.

European aviation expertise in the Community nowadays is incorporated in EASA. The Agency has legal personality and hence its own budget covered by specific financial rules. The annual programme of work shall aim to promote the continuous improvement of European aviation safety and comply with the objectives, mandates and tasks of the Agency, as set out in Regulation (EC) No 216/2008 (Basic Regulation) of the European Parliament and of the Council of 20 February 2008.[45]

Regulation (EC) No 216/2008 lays down the sources of revenue of the Agency, including fees and charges, levied by the Agency. Article 59, Budget:

1. The revenues of the Agency shall consist of:
   (a) A contribution from the Community;
   (b) A contribution from any European third country with which the Community has concluded agreements as referred to in Article 66;
   (c) The fees paid by applicants for, and holders of, certificates and approvals issued by the Agency;
   (d) Charges for publications, training and any other services provided by the Agency; and
   (e) Any voluntary financial contribution from member States, third countries or other entities, provided such a contribution does not comprise the independence and impartiality of the Agency.

45 *Official Journal of the European Union L* 79/1, 19 March 2008, at 1. *See also* (the same) Regulation (EC) No 216/2008, Article 56 Annual programme of work.

In accordance with Articles 9 and 23 of the Regulation (EC) No 216/2008, which entered into force on 8 April 2008, the Agency is given a number of new tasks with regard to operators of aircraft registered in a third country, or registered in a member State, which has delegated their regulatory safety oversight to a third country, and used by a third-country operator into, within or out of the Community.

As a matter of fact, EASA had already been established on 15 July 2002 by Regulation (EC) No 1592/2002 of the European Parliament and of the Council of 15 July 2002 on common rules in the field of civil aviation and establishing a European Aviation Safety Agency, and given legal personality and implementation powers. This Regulation was repealed by the Basic Regulation on this subject in 2008, as mentioned before.

The Basic Regulation is laying down the framework, criteria and objectives for EASA's new responsibilities. EASA consequently fills in the more technical details by drafting the associated Implementation Rules (IR), accompanied by non-binding Guidance Material (GM) and Acceptable Means of Compliance (AMC).

It should be noted that ICAO contracting States, which are also member States of EASA, will always maintain their individual responsibility for regulating their aviation industries and for implementing (aviation related) EU legislation, *e.g.* regarding pilot licensing, aircraft operations and inspection of aircraft used by third-country operators, taking into account the requirements of the Chicago Convention and the minimum standards established by ICAO.

EU member States may implement (more detailed) national rules and modify them as necessary, but they must abstain from taking measures that would affect the exercise of the established Community (EASA) competence.

In the coming years EASA gradually will issue rules and regulations, which will replace part of the national regulatory of the member States. The first extension, planned in 2012, concerns regulation regarding Air Crew and Air Operations, followed by Aerodromes and ATC.[46]

On the basis of its technical expertise EASA supports the European Commission with the implementation of the EC SAFA programme in the Community and as such ensures that appropriate follow-up measures are initiated based upon the SAFA inspections that have been performed. One of the main tasks of EASA currently includes authorization of third-country operators and the development of close working relationships with counterpart organizations across the world, including ICAO, the FAA in the United States, and the national civil aviation authorities of Canada, Brazil, Israel, the People's Republic of China and Russia.[47]

46 *See also* Commission Regulation (EU) No 1178/2011 of 3 November 2011 laying down technical requirements and administrative procedures to civil aviation aircrew pursuant to Regulation (EC) No 216/2008 of the European Parliament and of the Council, *Official Journal of the European Union L* 311/1, 25 November 2011.

47 <www.easa.eu.int/rulemaking/international-cooperation.php>.

Working arrangements between EASA and these and other organizations are aimed at harmonizing standards and promoting best practice in aviation safety and exchange of information on a global scale.[48]

International organizations like the European Union do not have the right to vote in the ICAO Assembly. All contracting States have equal right to be represented in the Assembly, and each contracting State is entitled to one vote.

EASA, as a European partnership institution and a key technical body advising the European Commission on all aviation safety matters, has a direct interest in remaining informed of the developments within the ICAO Assembly, and actively trying to influence this notable decision-making forum, in synergy with the European institutions, such as the legislative European Commission and other administrative and technical organizations (*e.g.* EUROCONTROL) within the European region.

At the 37th Assembly Session an EASA Internal Working Group prepared and presented 13 Working Papers to the Assembly covering issues that included the Next Generation of Aviation Professionals, promoting Regional Safety Oversight Organizations and transition to Continuous Monitoring Approach. Such papers aim to promote European views on many technical matters in which EASA, as one European voice, will play a central role, through its rulemaking, certification and approval and standardization activities.[49]

1.8 Concluding Remarks

The 18 Annexes to the Chicago Convention contain Standards and Recommended Practices. These SARPs, universally accepted, cover all operational and technical aspects of aviation, including the many facilities and services required in support of the international aircraft operations. Collectively, they form the basis for the safe and orderly development of international civil aviation.

When properly implemented and applied, they are essential for safe and efficient air transportation around the world. Without such uniform rules and procedures, aviation would be at best chaotic and at worst unsafe.

> Civil aviation is extremely safe. When accidents occur, however, it is often because the rules and procedures outlined in the SARPs and related guidance material developed by ICAO, are not adhered to properly.

48 Working Arrangements between EASA and several Civil Aviation Authorities (CAAs) of ECAC-non-EU member States on collection and exchange of information on the safety of aircraft using Community airports and the airports of the relevant country. Working Arrangements between EASA and several Civil Aviation Authorities of non-EU member States in the framework of the transition of the Joint Aviation Authorities.

49 EASA at the 37th ICAO Assembly, EASA News, January 2011.

> Compliance with ICAO SARPs will save lives and will contribute substantially to ensuring the safety of our skies in the 21st century.[50]

Safety oversight forms the part of the safety regulatory process dedicated to ensuring that applicable safety regulatory requirements are met, and to the monitoring of the safe provision of services. Unless prescribed otherwise by international legislation, safety oversight in aviation is a national responsibility, based on the sovereignty of a State (Article 37 of the Chicago Convention). ICAO Doc 9735 (Safety Oversight Audit Manual) states about sovereignty: Every State has complete and exclusive sovereignty over the airspace above its territory (Customary international law crystallized in the 1919 Paris Convention on the Regulation of Aerial Navigation and Article 1 of the 1944 Chicago Convention).

> Every State has, to the exclusion of all other States, the unilateral and absolute right to permit or deny entry into the area recognized as its territory and similar right to control all movements within such territory. (*Source*: Prof. John Cobb Cooper)

Accordingly, ICAO respects a contracting State's responsibility and authority for aviation safety oversight including its decision-making powers with respect to implementing corrective actions related to ICAO safety oversight audit findings. About universality: The mandate given to ICAO by the Assembly, through Resolution A32-11, is to perform safety oversight audits in all contracting States. Safety oversight audits shall, therefore, be conducted in accordance with an audit programme established by ICAO and accepted by the State to be audited.

International aviation-related regulations, standards, procedures and organization must be compatible and uniform to be able to equally enforce the implementation and to monitor compliance of the safety requirements throughout the international aviation world, regardless of any territorial boundaries.[51]

Comprehensive monitoring of compliance with safety rules and procedures by air carriers and compliance by States with legal binding international standards and technical regulations is of fundamental importance to verify the status of implementation. With its top-down universal audit programme called USOAP, ICAO introduced regular, mandatory, systematic and harmonized safety audits to determine the capability of its contracting States for safety oversight. In this respect ICAO assists contracting States in identifying and correcting deficiencies in the implementation of SARPs, which in fact are

50 Dr. A. Kotaite, President of the Council of ICAO, on the occasion of the 2000th edn (Montreal, 7 December 2000) of the International Civil Aviation Day, celebrated annually to mark the establishment of the ICAO on 7 December 1944.

51 *See also* J. Huang, *Aviation Safety and ICAO* (Ph. D. dissertation, Leiden University, Leiden), Kluwer Law International, Alphen aan den Rijn 2009, at 49-50.

continuously revised by ICAO, so that regulations keep pace with the rapid technological and operational transformation of civil aviation.

Another safety assessment program with a top-down approach is FAA's IASA. However, this programme is initiated by a single nation, the United States. The FAA meets with the foreign CAA, responsible for providing the safety oversight to its national carriers, which are proposing or having existing air services to the United States, reviews pertinent records and meets with officials of the subject foreign air carriers. The FAA assesses and analyses all information to determine whether the foreign CAA complies with ICAO standards regarding the oversight provided to the air carriers under its authority, and gives notice about a recommended course of action to the U.S. Department of Transport.

Regarding the two existing bottom-up assessment programmes, IOSA is focusing on air carrier safety using internationally recognized quality audit principles, and is designed so that audits are conducted in a standardized and consistent manner.

The SAFA programme, being a regional initiative, provides for inspections of foreign aircraft by the EU member States for the purpose of promoting the safety of international civil aviation. Air carriers unable to rectify shortcomings identified during these SAFA inspections and carriers operating under an inadequate system for regulatory oversight could be subject to an operating ban within the European Community. Bans and operational restrictions are only imposed on the basis of evidence of violation of objective and transparent criteria, criteria that focus on the assessment results of checks carried out at European airports.

2 International Legal Framework Regarding Airworthiness, Licensing and Safety Assessment

2.1 Introduction

Civil aviation, undoubtedly, is globally recognized as one of the most convenient, fastest and safest modes of travel. One reason for the rapid growth of today's air transport is, in no small measure, major improvements in aviation safety, as such safety is constantly being monitored, both domestically and internationally. However, all over the world, civil aviation is increasingly becoming aware of many hidden safety weaknesses. Such discoveries are also accelerating; the more one discovers, the more questions are being asked about specific safety aspects that previously were generalized. Increasing concern has been expressed for some time about levels of aviation safety in various parts of the world.

This prompted ICAO and the Government of the United States to establish oversight and assistance programmes in order to be able to identify possible deficiencies and to take corrective measures to ensure the safety of air travellers as well as people on the ground.

The European national civil aviation authorities themselves realized that the European Community and neighbouring countries also had to act in this particular field, resulting in the (voluntary) ECAC SAFA programme (1996), as a first step in addition to the global ICAO audit programme (SOP 1995, followed by USOAP 1999) and the U.S. national safety assessment program (IASA 1992).

These specific programmes are aimed at ensuring compliance with international safety standards in civil aviation. The need to take comprehensive action with regard to foreign air carriers and aircraft arises from the failure of an increasing number of countries to meet their obligations concerning the implementation and enforcement of these international standards.

The Chicago Convention and its Annexes allocate the responsibility for aviation safety to individual States, most of which are ICAO contracting States.[1] Each State bears responsibility for the safety design and continuing airworthiness of aircraft, in addition to safe operations, safe air traffic flow within its own national airspace and airport infrastructure. However, aviation safety deficiencies may arise if these obligations are not fully respected by States.

States, other than the responsible State, can only conduct specific surveillance activities, principally involving inspection of required documents and the physical condition of aircraft.

1 Article 31 of the Chicago Convention.

Most civil aviation authorities of ICAO contracting States (190 in 2011) issue specific airworthiness certificates and licences under requirements at least equal to the applicable minimum standards laid down in Annex 1 (Personnel Licensing) and Annex 8 (Airworthiness of Aircraft) to the Chicago Convention.[2]

Under Article 31 of this Convention every aircraft engaged in international navigation shall be provided with a Certificate of Airworthiness issued or rendered valid by the State in which the aircraft is registered. This text remarkably resembles the provisions of Article 11 of the Convention Relating to the Regulation of Aerial Navigation signed on 13 October 1919 in Paris. This agreement is largely based on the drafts of the 1910 Paris International Air Navigation Conference.

Article 11 states:

> Every aircraft engaged in international navigation shall, in accordance with the conditions laid down in Annex B, be provided with a Certificate of Airworthiness issued or rendered valid by the State whose nationality it possesses.

The nationality of aircraft is defined as follows:

> Aircraft have the characteristic of nationality similar to that developed in maritime law applicable to ships. Thus aircraft have normally a special relationship to a particular State which is entitled to make effective the privileges to which such aircraft may be entitled and such State is also reciprocally responsible for the international good conduct of such aircraft. (Prof. John Cobb Cooper)

Article 32 of the Chicago Convention stipulates that pilots and other members of the operating crew, engaged in international navigation, shall be provided with certificates of competence and licences issued or rendered valid by the State of Registry. Experience has shown, *inter alia*, by audits of the Universal Safety Oversight Audit Programme conducted by ICAO, that a State authority cannot reasonably assume, without verification, that the condition for recognition, stated in Article 33, is actually being met by another State.[3]

Article 33 states that recognition of such certificates and licences by other ICAO contracting States is fundamental for international civil aviation, provided that the requirements under which such documents were issued or rendered valid by the contracting State in which the aircraft is registered are equal to or above the minimum standards that may be established from time to time pursuant to the Chicago Convention.

2 ICAO Annex 1, 10th edn, 2006. ICAO Annex 8, 10th edn, April 2005, Amendment 99, May 2006.

3 ICAO Working Paper DGCA/06-WP/8 Directors General of Civil Aviation Conference on a Global Strategy for Aviation Safety, Montreal 20-22 March 2006, Theme 2: Improvement of Aviation Safety, 9 January 2006, at 2.

The harbinger of Article 33 is Article 13 of the aforementioned 1919 Paris Convention:

> Certificates of airworthiness and of competency and licenses issued or rendered valid by the State whose nationality the aircraft possesses, in accordance with the regulations established by Annex B and Annex E and hereafter by the International Commission for Air Navigation (ICAN), shall be recognized as valid by the other States.

Without such recognition, international flights would not be possible. Concerning the crucial recognition, each contracting State reserves the right to refuse to recognize, for the purpose of flight above its own territory, certificates of competency and licences granted to any of its nationals by another contracting State.[4]

The Chicago Convention supersedes other previous aeronautical agreements (Paris and Havana Conventions) as stated in Article 80:

> Each contracting State undertakes, immediately upon the coming into force of this Convention, to give notice of denunciation of the Convention relating to the Regulation of Aerial Navigation signed at Paris on October 13, 1919 or the Convention on Commercial Aviation signed at Havana on February 20, 1928, if it is a party to either. As between contracting States, this Convention supersedes the Conventions of Paris and Havana previously referred to.

For many years, States were generally confident that the other ICAO contracting States were fulfilling their (minimum) obligations under the Chicago Convention and recognized their certificates and licences at face value. However, USOAP audits showed that certificates and licences were being issued or rendered valid by a number of contracting States without the applicable standards being met.[5] As a result, recognition, and this is fundamental, can no longer be based solely on the principle of mutual confidence. An exception to the rule:

> While Article 6 of the Bermuda II Agreement should be seen as a provision having primarily precedent value, the Agreement of 1972 between the United Kingdom and the United States relating to the Reciprocal Acceptance of Airworthiness Certificates (HMSO, 1973, Cmnd 5321) is an agreement based on the highest degree of mutual confidence in each other's certification standards.
>
> Unlike the international agreements so far mentioned that deal with the recognition of certificates of airworthiness of complete aircraft for the purpose of international air service, the 1972 Agreement deals with recognition in the

4 Article 32(b) of the Chicago Convention.

5 ICAO Working Paper DGCA/06-WP/8, 9 January 2006, at 3.

context of export of aircraft, engines, propellers and other components. Its primary objective is to eliminate unnecessary testing of these products for the purpose of the issue of a Certificate of Airworthiness by the State of Registry and thus to eliminate a barrier to trade in aircraft.[6]

2.2 CERTIFICATES AND LICENCES

In order to enable a State to verify that relevant minimum standards are met when certificates and licences are issued or rendered valid and continue to be met, that State should have a process of recognition, based on harmonized criteria. Guidance material for the process of verification would be useful as a prerequisite to recognition and could include the following steps:

- identify the contracting State(s) responsible for the issuance or rendering valid the applicable certificates and licenses (State of Registry, State of the Operator and any other State(s)); this includes a review of any bilateral agreements, under Article 83 bis or otherwise, concerning the transfer of certain safety oversight functions and duties, as well as any lease agreements by the operator concerned;
- assess the level of safety oversight exercised by the contracting State(s) concerned; this includes a review of USOAP or any other safety oversight audit reports, in particular the States' effective implementation of their licensing and certification obligations and their surveillance obligations;
- assess whether the level and nature of aviation activity are consistent with the safety oversight capability of the contracting States concerned, *i.e.* the ability of the authorities responsible for safety oversight to implement and enforce the relevant safety standards, taking into account whether or not the Air Operator Certificate (AOC) has been issued by the responsible authority in the State where the operational control of the air operator resides.[7]

Apart from Article 33, Articles 16 and 21 of the Chicago Convention are relevant to a State's safety oversight responsibility to ensure safety in the airspace above its territory, including the operation of aircraft of foreign operators.

Article 16 contains the right of State authorities to search aircraft of the other contracting States on landing or departure on their territories, and to inspect the certificates and other documents prescribed by the Chicago Convention.

Article 21 includes special rules concerning on-demand reporting of information regarding the registration and ownership of any particular aircraft registered in a certain

6 E. Denza, 'From Aerostats to DC-10s-Recognition of Certificates of Airworthiness', *in* A. Kean (Ed.), *Essays in Air Law*, Martinus Nijhoff Publishers, The Hague 1982, at 47-48.

7 ICAO Working Paper DGCA/06-WP/8, 9 January 2006, at 1, 2.

contracting State.[8] Specific provisions and guidance material pertaining to the responsibility of a State related to foreign air operations within its territory should be developed to complement the obligations in the relevant articles of the Chicago Convention.[9] Surveillance over aircraft operations is needed, not only over operations under the direct responsibility of a contracting State, but also over foreign aircraft within their own territory and will not be limited to aircraft engaged in scheduled air services, but will also apply to non-scheduled flights.[10]

The globalization and expansion of international civil aviation over the last few decades, enhanced by the tremendous advancement of aviation technology, have created complexities that were not envisioned in the Chicago Convention.

However, the ICAO Annexes have been kept up to date to reflect the changing environment and to facilitate the aviation safety oversight by States and regional organizations. They provide the minimum requirements with the aim of standardizing international air transport (licensing, aircraft operation and airworthiness) to ensure the highest level of safety and efficiency.[11]

Annex 1, Personnel Licensing, provides for Standards and Recommended Practices for the licensing of flight crew members (pilots, flight engineers and flight navigators), air traffic controllers, aeronautical station operators, maintenance technicians and flight dispatchers.

The essence of Annex 6, Operation of Aircraft, is that the operation of aircraft engaged in international air transport must be as standardized as possible to ensure the highest levels of safety and efficiency.

The purpose of this Annex is to contribute to the safety of international air navigation by providing criteria for safe operating practices, and to contribute to the efficiency and regularity of international air navigation by encouraging ICAO contracting States to facilitate the safe passage over their territories of commercial aircraft belonging to other countries that operate in conformity with these criteria. In this connection the concept of an Air Operator Certificate, including the standards concerning the safety obligations of an air operator, were introduced.

Annex 8, Airworthiness of Aircraft, includes broad standards that define, for application by the national airworthiness authorities, the minimum basis for the recognition by

8 Article 21 Chicago Convention: Each contracting State undertakes to supply to any other contracting State or to the ICAO, on demand, information concerning the registration and ownership of any particular aircraft registered in that State. In addition, each contracting State shall furnish reports to the ICAO, under such regulations as the latter may prescribe, giving such pertinent data as can be made available concerning the ownership and control of aircraft registered in that State and habitually engaged in international navigation. The data thus obtained by ICAO shall be made available by it on request to the other contracting States.

9 ICAO Working Paper DGCA/06-WP/8, 9 January 2006, at 3.

10 *See also* Regulation (EC) No 2111/2005, Article 10.2.

11 Annexes 2 (Rules of the Air), 5 (Units of Measurement), 7 (Aircraft Nationality and Registration Marks) and 8 (Airworthiness of Aircraft) contain only Standards (minimum requirements). The remaining 14 Annexes contain Standards and Recommended Practices.

States of Certificates of Airworthiness for the purpose of flight of aircraft of other States into and over their territories, thereby achieving, among other things, protection of other aircraft, third parties and property.

> Airworthiness is the ability of an aircraft or other airborne equipment or system to operate without significant hazard to flight and cabin crew, ground crew, passengers, cargo or mail (where relevant) or to the general public and property over which such airborne systems are flown.[12]

Title 49, United States Code (49 U.S.C.) Paragraph 44704 (c) and FAA Federal Aviation Regulation 14 CFR Part 21 Paragraph 21.183 (a), (b) and (c) state that the following two conditions are necessary for issuance of an airworthiness certificate in the FAA's airworthiness certification programme:

a. The aircraft must conform to its type design. Conformity to the type design is considered attained when the aircraft configuration and the engine, propeller, and articles installed are consistent with the drawings, specifications, and other data that are part of the Type Certificate. This includes any Supplemental Type Certificate (STC) and repairs and alterations incorporated into the aircraft.
b. The aircraft must be in a condition for safe operation. This refers to the condition of the aircraft relative to wear and deterioration, for example skin corrosion, window delamination/crazing, fluid leaks and tire wear.[13]

To facilitate the import and export of aircraft and the exchange of aircraft for lease, charter or interchange as well as to facilitate operations of aircraft in international air navigation, Article 33 of the Chicago Convention places on the State of Registry the burden of recognizing and rendering valid an airworthiness certificate issued by another contracting State, subject to the condition that the airworthiness requirements under which such a certificate is issued or rendered valid are equal to or above the minimum standards that may be established by ICAO from time to time pursuant to the Chicago Convention. These minimum standards are contained in Annex 8, the first edition of which was adopted by the ICAO Council on 1 March 1949.

It is recognized that ICAO standards would not replace national regulations and that national codes of airworthiness containing the full scope and extent of detail, considered necessary by individual States, would be required as the basis for the certification of individual aircraft.

Each State is free to develop its own comprehensive and detailed code of airworthiness or to select, adopt or accept a comprehensive code established by another contracting

12 UK Ministry of Defence.

13 *See also* FAA FAR 14 CFR Part 3 Paragraph 3.5 (a).

State.[14] However, the level of airworthiness required to be maintained by a national code is indicated by the broad standards of Annex 8 supplemented where necessary, by guidance material provided in Doc 9760 (ICAO Airworthiness Technical Manual). Nevertheless, these ICAO standards still have a number of shortcomings.

ICAO Standards and Recommended Practices, being the minimum required standards to participate in international air transport, are a compromise of almost all contracting States, and are thus often only the lowest agreeable safety rule.

> (. . .) While implementing of these Standards and Recommended Practices is the surest way to advance safety globally, the SARPs are not enforceable by ICAO, and it is up to each contracting State to implement them to the extent possible. Any State which finds it impracticable to comply in all respects with a standard is required, under Article 38 of the Chicago Convention, to give immediate notification by filing a difference with ICAO. ICAO then notifies other States of the difference.[15]

There might still be a misconception concerning the legal status of the ICAO Standards and Recommended Practices, contained in the 18 Annexes to the Chicago Convention. The question about the binding force of ICAO standards is governed by the dichotomy whether the Annexes are to be considered inseparable parts of the Convention resulting in the full force effect of a formal source of adopted international law equivalent to an international convention, whether general or particular, establishing rules expressly recognized by the signatory States or a multipartite *per se* law-creating treaty, or just unenforceable, non-binding guidance material.[16]

> The legal obligation of States to comply with the standards adopted by the Council of ICAO under Articles 37, 54 and 90 of the Chicago Convention is not unconditional and the scope of the obligations is limited to the 'highest practicable degree'. Subject to the general requirement of good faith in the implementation of international obligation, only the State concerned can be the ultimate judge of what is 'practicable' in the given circumstances. However, if a State finds it impracticable to comply in all respects with any international

14 Annex 8 to the Chicago Convention, Airworthiness of Aircraft, Introduction.

15 A. Kotaite, 'Universal Audit Programme Conducted under Auspices of ICAO Would Offer Many Benefits', *ICAO Journal*, Vol. 53, No. 4, 1998, at 5.

16 Treaty is the generic term used to embrace convention, agreement, arrangement, protocol, and exchange of notes. International law does not distinguish between agreements identified as treaties and other agreements. See also the definition of a treaty by the 1969 Vienna Convention on the Law of Treaties: An international agreement concluded between States in written form and governed by international law, whether embodied in a single instrument or in two or more related instruments and whatever its particular designation.

> standard, it has an unconditional legal duty, under Article 38 of the Chicago Convention, to give immediate notification to ICAO and the Council of ICAO is obliged to make immediate notification to all other States of the difference that exists between an international standard and the corresponding national practice of that State.[17]

There is no effective global control mechanism established yet to ensure that these requirements are legally met, except the (in the past) voluntary and confidential assessment/audit programme established by ICAO (USOAP), which might be considered only as an initial step towards such a global monitoring mechanism.[18] The issue of worldwide enforcement of aviation safety standards cannot be disregarded. These internationally agreed safety standards aim at uniformity while permitting certain flexibility and this will undoubtedly enhance aviation safety. However, there is still no full implementation coverage with respect to SARPs by the ICAO member States, nor complete and adequate notification of differences, thus deteriorating the effectiveness of these regulations and rules of conduct.

Contracting States shall comply with the existing framework of their functions and responsibilities as well as their obligations that they have legally assumed under the Chicago Convention. However, without exception, every State has the right and responsibility to control their own safety codes to regulate the entire scope of their aviation industry as well as the right to deviate (differences) from the SARPs, which makes it all less similar, let alone uniform.[19]

2.3 SAFA, EASA AND THE EUROPEAN COMMISSION

Air transport in the European Community has been progressively liberalized since 1988, and the Community market in the civil aviation sector has evolved from a market based on bilateral agreements towards a genuine open market based on the principles of the European Treaty, without virtually any competition.

This process required the establishment and uniform application of common rules in the field of aviation safety as well as a harmonized safety oversight programme. This oversight programme is applicable not only to Community air carriers, but to a large extent to third-country air carriers, as reflected in the questionable airworthy condition of their aircraft operating within the Community.

17 M. Milde, 'Enforcement of Aviation Safety Standards – Problems of Safety Oversight', 45 *Zeitschrift für Luft- und Weltraumrecht* 1996, at 5.

18 ICAO created the ICAO Universal Safety Oversight Audit Programme in January 1999, focused on State oversight capability and compliance with ICAO Annexes.

19 States have the right (and corresponding duty) to notify differences between their national practices and internationally adopted standards (*See also* Article 38 Chicago Convention).

Safety deficiencies worldwide and the inability of some countries to meet their obligations concerning implementation and enforcement of international safety standards have an unacceptable impact on the European Union.

EU air carriers operate globally, and EU citizens travel widely all over the world and constitute an important percentage of all air travellers. A multitude of commercial flights are operated by third-country aircraft on a daily basis within the European airspace. Many EU citizens are travelling to and from Community airports on board these aircraft. Safety hazards of third-country aircraft have, given the international dominance of civil aviation, an adverse impact on air safety in the European region.

On 15 February 1996, the European Parliament adopted a Resolution calling on the European Commission to take a number of measures to improve aviation safety and to protect the European traveller. In addition, the Council of Transport Ministers proposed on 11 March 1996 that the European Commission establish a High Level Group of Experts to examine the various relevant issues in the context of aviation safety.[20] In 1996 the European Commission issued a Communication to the Council and the European Parliament concerning a Community Aviation Safety Improvement Strategy, clearly stating that safety may be effectively enhanced by ensuring that aircraft comply fully with internationally agreed safety standards contained in the Annexes to the Chicago Convention.[21]

This strategy was based on two pillars. The first pillar was a recommendation to work intensively to maintain and further improve civil aviation safety in Europe, particularly through an institutionalized body that should deal with aviation safety within Europe and beyond.

This initiative has eventually been developed with the approval of the European Commission on 27 September 2000 of a proposal regarding an EU Regulation of the European Parliament and of the Council on establishing common rules in the field of civil aviation and creating a European Aviation Safety Agency (COM (2000) 595).

The second pillar consisted of assisting foreign countries in safety oversight and implementing and enforcing international aviation safety standards and monitoring compliance with those standards by these countries to enhance overall safety regarding civil air transport. In the Community it is necessary for all member States to strictly monitor compliance within their respective airspace and at their airports, because some of these foreign countries occasionally fail in implementing and enforcing those standards.

Therefore the rules and procedures for ramp inspections, including grounding of an aircraft, the formal prohibition of an aircraft to depart from an airport and the taking of such steps as are necessary to detain it, should be harmonized to ensure consistent

20 The High Level Group of Experts consisted of representatives of all relevant competent authorities concerned, in particular the Directors General of Civil Aviation of the member States, ECAC and JAA, under the chairmanship of the European Commission.

21 Community Aviation Safety Improvement Strategy, SEC (96) 1083 final, 12 June 1996.

effectiveness at all airports throughout the Community, which in turn would also reduce the selective use of certain destination airports by third-country aircraft trying to avoid proper control.[22]

A harmonized approach to the effective enforcement of international safety standards by the member States prevents distortion of competition. A common stance in the relation with third countries, not complying with those standards, will strengthen the position of member States.

The objective of Directive 2004/36/EC, as stated in Article 1 of the Proposal for a Directive of the European Parliament and of the Council on the safety of third countries aircraft using Community airports, reads: The purpose of this Directive shall be to contribute to the improvement of air safety by ensuring that:

- information is collected and disseminated so that sufficient evidence can be established to decide on measures required to ensure the safety of the travelling public as well as of people on the ground;
- third-country aircraft, their operation and crew are inspected whenever there is reasonable suspicion that international safety standards are not being met and such aircraft are grounded if this is necessary to ensure immediate safety;
- appropriate measures for rectification of identified shortcomings are decided and implemented.

Within the framework of the Community's overall strategy to establish and maintain a high uniform level of civil aviation safety in Europe, this Directive introduces a harmonized approach to the effective enforcement of international safety standards within the Community by harmonizing the rules and procedures for ramp inspections on, in principle, third-country aircraft landing at airports located in the member States. An exception to the rule is made for certain aircraft. State aircraft, as defined in Article 3 (Civil and state aircraft) of the Chicago Convention,[23] and aircraft with a maximum take-off mass (MTOM) <5,700 kg, not engaged in commercial air transport, shall be excluded from the scope of this Directive. However, this Directive shall be without prejudice to the right of member States, with due regard to Community and international law, to carry out inspection not covered by this Directive, and to ground, ban, or impose conditions on, any aircraft landing at their airports.[24]

With respect to airport ramp inspections, each member State shall put in place the appropriate means to ensure that third-country aircraft, landing at any of its airports open to international air traffic and suspected of non-compliance with international safety

22 COM (2002) 8 final, Proposal for a Directive of the European Parliament and of the Council on the safety of third countries aircraft using Community airports, 14 January 2002.

23 Article 3(b): Aircraft used in military, customs and police services shall be deemed to be state aircraft.

24 Directive 2004/36/EC of the European Parliament and of the Council of 21 April 2004 on the safety of third-country aircraft using Community airports, *Official Journal of the European Union L* 143/76, 30 April 2004, Article 1.

requirements, shall be subject to those inspections (Proposal COM (2002) 8 final, Article 5 Ramp Inspection, embodied in Article 4 of Directive 2004/36/EC).

In implementing such procedures, particular attention shall be given by the competent authority to aircraft:

- where information has been received indicating poor maintenance condition or obvious damage or defects;
- which have been reported as performing abnormal manoeuvres since entering the airspace of a member State, such as to give rise to serious safety concerns;
- in respect of which a previous ramp inspection has revealed deficiencies which give rise to serious concern that the aircraft does not comply with international standards and where the member State is concerned that the defects may not have been corrected;
- there is evidence that the competent authorities of the State of Registry may not be exercising proper safety oversight; or
- where information collected under COM (2002) 8 final, Article 4 (Collection of Information, embodied in Article 3 Directive 2004/36/EC) gives cause for concern about the operator or where a previous ramp inspection of an aircraft used by the same operator has revealed deficiencies.

Collection of relevant information, as stated in the Proposal's objective, will be done in a way that member States shall put in place a mechanism to collect any information deemed useful for the fulfilment of the applicability of this Directive, including:

(a) important safety information accessible, in particular through:
   - pilot reports,
   - maintenance organization reports,
   - incident reports,
   - other organizations, independent from the competent authorities of the member States,
   - complaints;

(b) information on action taken subsequent to a ramp inspection, such as:
   - aircraft grounded,
   - aircraft or operator banned from the member State concerned,
   - corrective action required,
   - contacts with the operator's competent authority;

(c) follow-up information concerning the operator, such as:
   - corrective action implemented,
   - recurrence of discrepancy.

Safety related information gathered in each member State should be made available to all member States and the European Commission, in order to enhance monitoring and supervising of compliance with international safety standards by third-country civil aviation authorities in the most efficient way.

The sensitive nature of this information is such that the only way to ensure the gathering of such information is by ensuring its proper usage and the confidentiality of its sources without prejudicing the right of the citizens of the European Union to be informed of the efforts made to ensure their safety, and of the most serious cases affecting air safety. Therefore the competent authorities of the member States shall participate in a mutual exchange of safety-related information. This information shall be used solely for the purpose of SAFA Directive (2004/36/EC) and its access shall be limited to the participating, competent authorities and the European Commission.

The management role is fulfilled by EASA and consists of collecting data from the member States relevant to the safety information on aircraft using Community airports, developing, maintaining and providing continuous updating of a (computerized) centralized databank containing all compulsory, collected relevant data.

Further management tasks include providing necessary changes and enhancements to the database application, analyzing the centralized database information and other relevant information concerning the safety of aircraft and of operators and maintaining liaison with other European institutions and bodies, international organizations and national civil aviation authorities on information exchange.[25]

2.4 Community List

Where non-compliance with international safety standards is clearly hazardous to safety, the competent authority performing the ramp inspection shall ground the aircraft until the hazard is removed. In the event that an aircraft is grounded, the competent authority of the member State where the inspection took place shall immediately inform the competent authorities of the operator concerned and of the State of Registry of the aircraft involved. If the hazard is such that the aircraft may be safely flown without revenue passengers or cargo, the competent authority of the member State where the inspection took place shall, in coordination with the State responsible for the operation of the respective aircraft, prescribe the necessary conditions under which the aircraft could be released once-only to fly in a safe manner, explicitly without revenue passengers or cargo, to an airport at which the deficiencies could be corrected or repaired to rectify dangerous situations, and inform the States which will be overflown by such aircraft.

25 Commission Regulation (EC) No 768/2006 of 19 May 2006 implementing Directive 2004/36/EC of the European Parliament and of the Council as regards the collection and exchange of information on the safety of aircraft using Community airports and the management of the information system, *Official Journal of the European Union L* 134/16, 20 May 2006.

If the deficiencies affect the validity of the Certificate of Airworthiness for the respective aircraft, the grounding of that aircraft may be lifted only if the operator obtains permission from the State or several States that will be overflown by that particular flight.[26]

If corrected measures are not forthcoming and not satisfactorily implemented, the member State concerned may decide to totally ban the third-country operator or impose restrictive conditions on the operations to airports of that member State, pending the adoption of satisfactory arrangements for corrective measures by the competent authorities of that specific third country. In that case the member State in question shall notify the European Commission of the measures taken.

In addition, the European Commission may identify the Community-wide implications of the perceived safety hazard and may take any measures deemed necessary, including the extension to the Community of the measures taken by the member State concerned.

With respect to operating bans regarding (third-country) air carriers, Regulation (EC) No 2111/2005 of the European Parliament and of the Council of 14 December 2005 on the establishment of a Community list of air carriers subject to an operating ban within the Community and on informing air transport passengers of the identity of the operating air carrier has been adopted.[27]

This Regulation imposes a duty on EASA to communicate all information that may be relevant for the updating of the Community list of air carriers which, for obvious safety reasons, are subject to an operating ban within the Community.

If the Agency refuses to grant an authorization to an air carrier under the terms of this Regulation, it should pass on to the European Commission all relevant information on which this refusal was based, so that the name of the air carrier may be entered, if applicable, on that particular list.[28] The Regulation provides for a decision-making process whereby an air carrier may be banned from European airspace, solely for safety reasons.

The European Commission (Mobility and Transport) may decide, in accordance with this Regulation and based on various other sources of information, upon an operating ban within the Community.

Regulation (EC) No 2111/2005 has been amended by Commission Regulation (EC) 474/2006 of 22 March 2006, establishing the Community list of air carriers that are subject to an operating ban within the Community, as referred to in Chapter II of the original 2005 Regulation (2111/2005).[29]

26 COM (2002) 8 final, Article 8 (Grounding of aircraft), embodied in Directive 2004/36/EC, Article 7.

27 Regulation (EC) No 2111/2005 of the European Parliament and of the Council of 14 December 2005 on the establishment of a Community list of air carriers subject to an operating ban within the Community and on informing air transport passengers of the identity of the operating air carrier, and repealing Article 9 of Directive 2004/36/EC, *Official Journal of the European Union* L 344, 27 December 2005, at 15.

28 Regulation (EC) No 216/2008, (14).

29 Chapter II, Article 4 Updating of the Community list.

This 2006 Regulation in turn has been amended by Commission Regulation (EC) No 1071/2010 of 22 November 2010 (amending Regulation (EC) No 474/2006 establishing the Community list of air carriers that are subject to an operating ban within the Community). This Regulation was adopted following an agreement on the main operational conclusions presented by EASA and the European Commission in front of the Air Safety Committee in the course of the meeting of the European SAFA Steering Group (ESSG) held in Vienna on 28 and 29 October 2010.[30]

Amendments (updates) of the list are necessary because of frequent mutations. So far these amendments have basically consisted of continuous failure in providing complete and current operational specifications concerning airworthiness of the fleet of certain third-country air carriers.

Some third countries initially failed to demonstrate effectively discharging their responsibilities regarding safety oversight of certified air carriers, resulting in significant safety concerns about aircraft operations, certification and supervision of the relevant civil aviation authorities (based on ICAO USOAP audits performed in the countries involved).

2.5 The New Basic Regulation

Regulation (EC) No 216/2008 (Basic Regulation), amended by Commission Regulation (EC) No 690/2009 of 30 July 2009,[31] and Regulation (EC) No 1108/2009 of the European Parliament and of the Council of 21 October 2009,[32] sets common rules in the field of civil aviation and establishes a European Aviation Safety Agency as well as an appropriate and comprehensive framework for the definition and implementation of common administrative procedures and technical requirements in the field of civil aviation.

This Basic Regulation also establishes an appropriate and comprehensive legal framework for the safety of third-country aircraft using Community airports.[33] The objectives of the Basic Regulation are more diverse in relation to the SAFA Directive, although the premise is identical:

1. The principal objective of this Basic Regulation is to establish and maintain a high uniform level of civil aviation safety in Europe.

30 The European Air Safety Committee consists of technical air safety experts from all the EU member States, including the States of Iceland, Norway and Switzerland, which however have no voting rights.

31 Article 6(1) of Regulation (EC) No 216/2008 is replaced by the following: Products, parts and appliances shall comply with the environmental protection requirements contained in Amendment 9 of Volume I and in Amendment 6 of Volume II of Annex 16 to the Chicago Convention as applicable on 20 November 2008, except for the Appendices to Annex 16.

32 This Regulation amended Regulation (EC) No 216/2008 in the field of aerodromes, air traffic management and air navigation services.

33 Directive 2004/36/EC should therefore be repealed in due course and without prejudice to the implementation measures (Articles 3, 4 and 5) on collection of information, ramp inspection and exchange of information (32-34, Regulation (EC) No 216/2008).

2. Additional objectives are, in the fields covered by this Basic Regulation, as follows:
   (a) to ensure a high uniform level of environmental protection;
   (b) to facilitate the free movements of goods, persons and services;
   (c) to promote cost-efficiency in the regulatory and certification processes and to avoid duplication at national and European levels;
   (d) to assist member States in fulfilling their obligations under the Chicago Convention, by providing a basis for a common interpretation and uniform implementation of its provisions, and by ensuring that its provisions are duly taken into account in this Regulation and in the rules drawn up for its implementation;
   (e) to promote Community views regarding civil aviation safety standards and rules throughout the world by establishing appropriate cooperation with third countries and international organizations;
   (f) to provide a level playing field for all actors in the internal aviation market.[34]

Where the scope of the current SAFA Directive is limited to the introduction of a harmonized approach to the effective enforcement of international safety standards within the Community by harmonizing the rules and procedures for ramp inspections of third-country aircraft landing at airports located in the member States, the Basic Regulation (total system approach) shall apply to design, production, maintenance and operation of aeronautical products, parts, appliances, aircraft, aerodromes, aerodrome equipment, systems and constituents for Air Traffic Management and or Air Navigation Services (ATM/ANS) as well as to personnel and organizations involved in these specific areas, being all non-military.[35]

As a matter of fact, in this perspective it is a misconception that aviation safety audits are for air operators only. With the present knowledge that some aviation accidents are caused by failures that may be far removed from the actual direct accidents, all aspects of civil aviation are subject to full safety evaluation.

If, for example, aerodrome operations are not conducted in accordance with internationally established standards and procedures, air operations, and in fact ground operations, may be prone to cause accidents. During the safety audit or assessment process of one air operator it is possible that only one aerodrome link is exposed. Only an independent aerodrome audit could disclose all of its internal aviation safety weaknesses. The same is applicable to any aviation organization, including non-aviation-related organizations that interface with civil aviation.

34 Regulation (EC) No 216/2008, Article 2 Objectives.

35 Product shall mean an aircraft, engine or propeller. Parts and appliances shall mean any instrument, equipment, mechanism, part, apparatus, appurtenance, software or accessory, including communication equipment, that is used or intended to be used in operating or controlling an aircraft in flight; it shall include parts of an airframe, engine or propeller, or equipment used to manoeuvre the aircraft from the ground.

It is for this very reason that aviation safety auditing should not be limited to air operators only, but should be equally appropriate for all organizations and companies that have direct or indirect connections with civil aviation operations activities. These operations activities include aerodromes, ATC, State's civil aviation organizations/authorities, suppliers and shippers of dangerous goods, energy suppliers providing services to aerodromes, telecommunications companies responsible for providing teledata communications/telephone lines to and from the aerodromes or other vital aviation support installations.

To further improve aviation safety, besides direct aviation safety audits, other methods must be used to directly inform individuals/organizations involved. Major air carriers have methods to increase the knowledge base of their staff directly involved in aircraft operation, such as Crew Resource Management, aircraft safety equipment and adverse weather conditions training, seminars, Aircrew Reference Guides and dangerous goods handling.

The scope of the Basic Regulation, as stipulated in Article 1, was amended in 2009.[36] The established total system approach, based on the fact that the aviation system components like products, operators, crews, aerodromes, ATM/ANS on the ground and in the air are part of a single network, also streamlines the certification processes and reduces the burden on regulated persons and organizations.

Rules stated in Article 9 of this Basic Regulation are applicable to aircraft used by third-country operators into, within or out of the Community. These aircraft, referred to in Article 4(1)(d),[37] as well as their crew, their certificates and their operations, shall comply with applicable ICAO standards.

The functions of the Basic Regulation include the following:

a. It defines the scope of Community competence, creates EASA, and establishes its powers, functions and organizational structure;
b. It establishes the division of regulatory and executive powers among EASA, the European institutions, and EASA member States;
c. It establishes the powers and functions of the Management Board and the Executive Director of the Agency (EASA);
d. It delegates the necessary authority to the Executive Director to fulfil the regulatory and executive competences transferred to EASA;
e. It grants EASA inspectors the necessary powers and authority to exercise the oversight functions of the Agency. (With respect to the inspectors of the competent authorities

36 *See also* Regulation (EC) No 1108/2009 of the European Parliament and of the Council of 21 October 2009 amending Regulation (EC) No 216/2008 in the field of aerodromes, air traffic management and air navigation services (ATM/ANS) and repealing Directive 2006/23/EC, *Official Journal of the European Union L* 309/51, 24 November 2009.

37 Article 4(1)(d): Aircraft registered in a third country or registered in a member State which has delegated its regulatory oversight to a third country and used by a third country operator, into, within or out of the Community.

of the member States, it is their national laws that empower them as a part of the enforcement system of the member States. Nevertheless, some paragraphs of the Implementing Rules provide the competent authorities the right of access to the premises of certificate holders and those of their subcontractors);

f. It specifies the power to revoke or suspend certificates and approvals.

The member States, the European Commission and EASA shall cooperate with a view to ensuring compliance with the provisions of the Basic Regulation and its Implementing Rules. For the purposes of the implementation of paragraph 1, the member States shall, in addition to their oversight of certificates that they have issued, conduct investigations, including ramp inspections, and shall take any measure, including the grounding of aircraft, to prevent the continuation of an infringement.[38]

The Basic Regulation has given the EASA additional operational responsibilities. In the field of flight crew licensing, the Agency is responsible for the approval and oversight of pilot training organizations and aero-medical centres located outside the territory of the Community. In addition, it certifies flight simulation devices located in and outside the Community on request. In the field of operations, the Agency was given the power to determine corrective actions and disseminate information to react with undue delay to a problem affecting the safety of air operations. With respect to third-country operators, the Agency is competent for the authorization and oversight of commercial air operators, as well as for the authorization of third-country aircraft when these aircraft or their crews do not comply with ICAO standards.

With regard to third-country operators, this particular Basic Regulation specifically requires that:

- All third-country operators (their aircraft, the crew and the operations) comply with the ICAO standards. To the extent that there are no such standards, these aircraft and their operations shall comply with the Essential Requirements for airworthiness, pilot licensing and air operations as laid down in Annexes I, III and IV of the Basic Regulation, provided these requirements are not in conflict with the rights of third countries under international conventions (Article 9(1));
- Third-country operators engaged in commercial air operations receive an authorization issued by the Agency (Article 9(2)); and
- Third-country operators engaged in non-commercial air operations of complex motor-powered aircraft may be required to declare their capability and means of discharging the responsibilities associated with the operation of that aircraft to the Agency (Article 9(3)).

38 Regulation (EC) No 216/2008, Article 10, Oversight and enforcement.

39 Some delay is possible as a result of incomplete texts. Regulations regarding Aircrew (Commission Regulation (EU) No 1178/2011) shall become effective from 8 April 2012, while Air Operations have been postponed until at least October 2012.

New Implementing Rules and relevant Operations Specifications and guidance material contain detailed provisions on how to meet the above requirements of the Basic Regulation. These Implementing Rules will incorporate the SAFA Directive while maintaining the same principles and competencies of oversight by member States. However, the SAFA Directive and national rules of member States continue to apply until the date specified in the relevant Implementing Rules, but not later than 8 April 2012 (Article 70), when the provisions for third-country operators shall be adopted.[39]

2.6 Concluding Remarks

In a number of contracting States, several of the requirements that would signify an appropriately established civil aviation organization are unfortunately missing and, consequently, there is no effective system established for regulating and supervising civil aviation activities or adopting relevant regulations.

This phenomenon often emerges as a corollary of third-country aircraft safety assessments and safety oversight audits performed in those contracting States. Shortcomings in the field of licensing, airworthiness certification and continuation as well as safe air carrier operations, in conjunction with an increasing global air transport market, and the need to control these activities for the benefit of society, have led to the emergence of international safety assessment and audit programmes.

Responsibility of safety oversight rests with each sovereign contracting State. The State's safety oversight capability is not limited to aircraft airworthiness certification and licensing of air crew, but encompasses all activities and actions to be taken regarding continuing airworthiness and recurrent monitoring of air crew competency. Safety oversight is a rather complicated activity owing to the steady and increasing international character of air carrier operations and the increasing movement of the operational bases of civil aircraft across national boundaries, especially in the European Community.

Another complicating feature nowadays is the fact that a substantial part of the fleet of the air carriers throughout the world is leased, resulting in the circumstance that the State of Registry may be different from the State of the Operator, which will render an effective safety oversight by the State of Registry difficult, if not impossible.

Air carriers are responsible for safe aircraft operations. Third-country aircraft, in particular, are subject to SAFA inspections by member States of the European Union. Experience has shown that third-country air carriers using Community airports do not always comply with international minimum safety standards, and this could place Community citizens travelling with these carriers or living in the vicinity of Community airports in danger. Ramp inspections, as part of the Community safety assessment programme, will be carried out in compliance with the international safety standards listed in the Annexes to the Chicago Convention to which all the member States of the European Union are party.

Community legislation and applicable ICAO SARPs form the foundation with respect to these inspections and safety data collection. EASA, a single specialized expert body, established within the institutional structure based on balance of powers and having legal and administrative as well as financial autonomy, is the coordinating European entity regarding the SAFA programme.

To facilitate the implementation of SAFA activity (inspection and data collection) rules in each member State, a coordination function is needed at a national level. Reports showing serious irregularities and safety deficiencies of air carriers may be relevant to immediately include the air carrier concerned on the so-called EU blacklist. These reports may originate from SAFA inspections, performed at airports within the European Union, operating bans imposed by non-EU countries, audit reports drawn up by ICAO following inspections of civil aviation authorities within the framework of USOAP, as well as accident-related information or other serious incident-related information.

Regarding the EU blacklist, the decision to include or remove a carrier is taken on the basis of the common safety criteria annexed to the EU ban Regulation (Regulation (EC) No 2111/2005 establishing a Community list of banned carriers). The rules establishing the EU blacklist of banned carriers apply to all commercial air carriers irrespective of their nationality – EU and non-EU ones.

For the overall safety of third-country aircraft using Community airports and relevant aviation safety rules, activities, requirements and procedures in the European Union, Regulation (EC) No 216/2008 (Basic Regulation) establishes an appropriate and comprehensive framework referred to as total system approach legislation.

This legislation provides for ensuring a high uniform level of protection of European citizens in civil aviation, through adoption of clearly defined common safety rules and measures, which in turn ensure that products, persons and organizations in the Community comply with such rules and adopted rules to protect the environment.

This framework should contribute to facilitating the free movement of goods, persons and organizations in the internal Community market. Appropriate oversight of third-country aircraft operating into, within or out of the territory of States where the Treaty (the Treaty on European Union and the Treaty establishing the European Community) applies will be accomplished according to the minimum standards ensuring safety of civil aviation and environmental protection relating thereto, as set by the Chicago Convention.

3 Airworthiness and Licensing

3.1 Airworthiness

A term used to describe both the legal and mechanical status of an aircraft with regard to its readiness for flight.

> A State must protect the safety of its nationals, even abroad: it is possible that in an aerostat taking off from one country there will be nationals of another State; or cargo belonging to citizens of one State may be carried by a foreign aerostat, and it is right that the State should know that they are not in danger on board, that they will arrive safely at their destination: the development of international commerce requires that air machines should be guaranteed to be airworthy. It is not just the safety of persons and goods on board that requires a prior permit, it is also the safety of persons and property in the countries below who are exposed to dangers and damage, particularly at the time of landing; and in this regard, foreign countries are just as interested as the country of the aerostat's nationality.[1]

The Wright brothers not only designed the first successful aircraft, they also built it and made the first controlled, powered and sustained heavier-than-air human flight on 17 December 1903.

Nowadays such versatility is inconceivable, indeed irresponsible. Most aircraft developed are the complex product of a team effort by a large number of aviation experts: designers, engineers and multiple technicians. In a time frame of enormous technological progress there is very little doubt that a newly developed aircraft will fly. Modern aircraft are not designed merely on simple assumptions or hunches, but according to known scientific laws, mathematical formulae, empiricism and profound computer-aided research. The result of all this is an aircraft with handling characteristics and performance similar, equal to or even better than computer calculations and wind tunnel tests.

The handling qualities and characteristics, performance requirements and procedures, all related to the airworthiness of a particular aircraft, are laid down in the flight manual,

1 Dr. Paul Auguste Joseph Fauchille (1858-1926). Fauchille's proposals regarding certificates of airworthiness, among other things, were endorsed by the Paris International Air Navigation Conference (1910) with the purpose of drafting an international convention to regulate air navigation.

subject to meeting at least the minimum standards established by ICAO and as a rule, apart from (notified) differences, implemented in relevant national regulations.[2]

Since the time that ICAO, the specialized technical agency of the United Nations (UN), was founded on 4 April 1947, the responsibility of this organization has been the creation, adoption and modernization of universally accepted 18 Annexes to the Chicago Convention, identified as SARPs, to ensure the safe, efficient and orderly evolution of international civil aviation.

An ICAO standard is defined as: any specification for physical characteristics, configuration, material, performance, personnel or procedure, the uniform application of which is recognized as necessary for the safety or regularity of international air navigation and to which contracting States will conform in accordance with the Chicago Convention.

In the event of impossibility of compliance, notification to the ICAO Council is compulsory under Article 38 to the Chicago Convention. ICAO standards do not preclude the development of national standards that may be more stringent than those contained in the Annexes.

An ICAO recommended practice is defined as: any specification for physical characteristics, configuration, material, performance, personnel or procedure, the uniform application of which is recognized as desirable but not essential in the interest of safety, regularity or efficiency of international air navigation, and to which contracting States will endeavour to conform in accordance with the Chicago Convention. States are invited to inform the ICAO Council of non-compliance.[3]

There are three Annexes that are directly connected with airworthiness:

> Annex 6. Operation of Aircraft. The purpose of Annex 6 is to contribute to the safety of international air navigation by providing criteria for safe operating practices, and to contribute to the efficiency and regularity of international air navigation by encouraging contracting States to facilitate the passage over their territories of commercial aircraft belonging to other countries that operate in conformity with these criteria. Annex 6 enumerates specifications to ensure the highest levels of safety and efficiency above a prescribed minimum in an as standardized as possible international air transport.
>
> Annex 8. Airworthiness of Aircraft. Annex 8 to the Chicago Convention includes broad standards which define, for application by the national airworthiness authorities, uniform procedures for certification and inspection of

2 Under the terms of Articles 37 and 38 of the Chicago Convention, the contracting States are obliged to bring their national regulations and procedures in line with the international SARPs and related procedures adopted and amended by ICAO, as far as possible and as practicable.

3 ICAO Air Navigation Bureau (ANB). Making an ICAO Standard, 1 November 2011.

aircraft considered as the minimum basis for the recognition by States of Certificates of Airworthiness for the purpose of flight of aircraft of other States into and over their territories, thereby achieving, among other things, protection of other aircraft, third parties and properties.[4]
The level of airworthiness required to be maintained by a national code is indicated by the broad standards of Annex 8 supplemented, where necessary, by applicable Guidance Material provided in ICAO's Airworthiness Technical Manual (Doc 9760).

Annex 13. Aircraft Accident and Incident Investigation. Actual and potential safety deficiencies, not least in the field of airworthiness, may lead to an accident or serious incident. This Annex provides for the international requirements for the investigation of aircraft accidents and incidents. The process of identifying and analysing the causal factors of such an occurrence is best accomplished through a properly conducted technical investigation in order to prevent similar future occurrences.[5] The investigation process includes the gathering, recording and analysis of all relevant information, the determination of the causes, formulating appropriate safety recommendations and the completion of the final report.

Other Standards and Recommended Practices, particularly focusing on human factors, capabilities, limitations and skills in international air transport, and thus inseparable from airworthiness in the broadest sense, are provided by Annex 1 (Personnel Licensing).

Today's aircraft operations are so diverse, demanding and complex that protection should be provided against the possibility, however rather remote, of total system breakdown due to either human error or failure of a system or a system component. The human being is still the vital link in the chain of aircraft operations and subject to proper training that is deemed necessary to minimize human error and provide capable, skilful, proficient and competent personnel at various jobs ranging from flight crew to maintenance technicians.

One of ICAO's main tasks in the field of personnel licensing is to foster the resolution of differences in licensing requirements and to ensure that international licensing standards are kept in line with current practices and probable future developments.

4 Introduction to Annex 8 Airworthiness of Aircraft.

5 SARPs for aircraft accident inquiries were first adopted by the Council on 11 April 1951, pursuant to Article 37 of the Chicago Convention, and were designated as Annex 13 to the Chicago Convention. Annex 13, Chapter 3, 3.1 Objective of the investigation: The sole objective of the investigation of an accident or incident shall be the prevention of accidents and incidents. It is not the purpose of this activity to apportion blame or liability.

Annex 1 is regularly amended to reflect the rapidly changing environment in which aviation professionals will be exposed to time pressure, increasing air traffic density and airspace congestion, highly complicated airport departure and arrival procedures and terminal area patterns and more sophisticated equipment and tools.

Annex 18, with strong links to legal aspects regarding airworthiness and safety assessment of aircraft, deals with the Safe Transport of Dangerous Goods by Air. Dangerous goods are certain types of material(s) needing special care and treatment because of a flammable, toxic or poisonous nature. When properly packed, stored, labelled, categorized, protected and treated, dangerous goods may be transported, although drastic transport restrictions with respect to uploading on passenger aircraft are applicable.

In general, Annex 18 sets down broad principles. However, one of the standards requires that dangerous goods be carried in accordance with the Technical Instructions for the Safe Transport of Dangerous Goods by Air. Technical Instructions consist of detailed requirements and provisions and have their own specific nature, *sui generis*, and are thus being classified between SARPs and Guidance Material.[6]

Contracting States are required by Annex 18 to have inspection and enforcement procedures to ensure that dangerous goods are being carried in compliance with international standards implemented in their national legislations under the provisions of the Chicago Convention, in order to safeguard aircraft and its occupants as well as personnel and residents on the ground. This system ensures governmental control over the carriage of dangerous goods by air and gives worldwide harmonization of safety standards, especially being reflected in random inspections of commercial aircraft like the SAFA inspections.

ICAO Annex 8 indicates that in the interest of safety an aircraft must be designed, constructed and operated in compliance with the appropriate airworthiness requirements of the State of Registry of that particular aircraft. Consequently, the aircraft will be issued with a Certificate of Airworthiness declaring that the aircraft is suitable to fly.[7]

Every aircraft must be in the possession of a Certificate of Airworthiness, certifying that the aircraft fulfils the appropriate safety requirements. However, such a certificate does not prove that the aircraft is airworthy at the start of each particular flight leg.

Any damage or deficiency, which could affect the safe operation of a flight, except for minor defects as defined by the Aircraft Operations Manual and Guidance Material like a Dispatch Deficiency Guide (DDG) or Minimum Equipment List (MEL) and subsequently

---

6 J. Huang, *Aviation Safety and ICAO* (Ph. D. dissertation, Leiden University, Leiden), Kluwer Law International, Alphen aan den Rijn 2009, at 64.

7 The Certificate of Airworthiness as used in the standards of Annex 8 is the Certificate of Airworthiness referred to in Article 31 of the Chicago Convention.

mentioned in a Hold Item List, must be repaired instantly to guarantee the state of airworthiness before resuming a flight.[8]

A Certificate of Airworthiness shall be renewed or shall remain valid, subject to the laws and regulations of the State of Registry, provided that this State shall require that the continuing airworthiness of the aircraft shall be determined by a periodical inspection at appropriate intervals having regard to lapse of time and type of service or, alternatively, by means of a system of inspection, approved by the State, that will produce at least an equivalent result.[9]

It is the sovereign right of the State of Registry to issue Certificates of Airworthiness subject to compliance with prescribed minimum requirements, *in casu* requirements equal to or above the minimum standards that may be established by ICAO from time to time pursuant to the Chicago Convention, and by ICAO resolutions amended as a consequence of the significant technological progress and more profound safety demands.

Other contracting States are required to recognize the certificates of contracting States of Registry, but only where the national standards of the certifying State are equal to or above the earlier mentioned minimum standards.[10]

The ICAO Annexes are using the following definitions related to airworthiness:

- State of Registry. The State on whose register the aircraft is entered.[11]
- State of Design. The State having jurisdiction over the organization responsible for the type design.

---

8 Deficiencies to the aircraft, cockpit and cabin are stated in the Hold Item List in the Aircraft (Technical) Maintenance Log. The aircraft may be released for a flight or series of flights with aircraft or cockpit system deficiencies if these deficiencies are stated in the Master Minimum Equipment List (MMEL), being the basis for the MEL or DDG (a description used by KLM until a few years ago), which provides or refers to conditions, limitations and procedures to permit operation with unserviceable items of equipment for a period of time until repairs must be accomplished at the earliest opportunity. The DDG/MEL additionally takes into consideration the operator's (KLM) aircraft configurations, route and other operational requirements for commercial air transports. All equipment not listed in the DDG/MEL, must be serviceable. The MEL is normally established by the aircraft operator and approved by the responsible State of the Operator, while the MMEL is established by the manufacturer as a baseline document for the operator to set up the MEL. An operator's MEL may differ in format from the MMEL, but cannot be less restrictive than the MMEL. The MMEL includes those items of equipment related to the airworthiness of the aircraft that may be unserviceable and yet maintain an acceptable level of safety by appropriate conditions, limitations and procedures. *See also* JAR-MMEL/MEL Section 1 Subpart B-MMEL definition: The MMEL is a document that lists the equipment which may be temporarily inoperative, subject to certain conditions, while maintaining an acceptable level of safety as intended in the applicable JAR or equivalent Requirement.

9 ICAO Annex 8 Part II Chapter 3, 3.2.3 and Chapter 4 (Continuing Airworthiness of Aircraft).

10 Article 33 of the Chicago Convention.

11 See Annex 7 to the Convention on International Civil Aviation, Aircraft Nationality and Registration Marks. In the case of the registration of aircraft of an international operating agency, other than on a national basis, the States constituting the agency are jointly and severally bound to assume the obligations which, under the Chicago Convention, attach to a State of Registry. See in this regard: ICAO Doc 9587 *Policy and Guidance Material on the Economic Regulation of International Air Transport*, 3rd edn, 2008.

- State of Manufacture. The State having jurisdiction over the organization for the final assembly of the aircraft.
- State of the Operator. The State in which the operator's principal place of business is located, or, if there is no such place of business, the operator's permanent residence.
- State of Occurrence. The State in the territory of which an accident or incident occurs.

The globalization and expansion of international civil aviation over the last decades, as well as the advancement of aerospace technology, have created complexities that were not envisioned in the Chicago Convention. The Annexes have been kept up to date to reflect the changing environment and facilitate the safety oversight by contracting States. In this view recognition of Certificates of Airworthiness is fundamental for international civil aviation. It is proposed that ICAO should develop provisions and guidance to promote uniformity in the recognition by States of Certificates of Airworthiness and licences as well as Air Operator Certificates as valid and in the surveillance of foreign aircraft operations within their territory.

Contracting States, as signatories to the Chicago Convention, have an obligation to ensure safety oversight of aircraft operations within their airspace, as ICAO has been called upon to play a leadership role in developing global strategies for the regulation and oversight of aviation safety through a system based on, although unfortunately, controversial mutual trust and recognition.

Article 33 of the Chicago Convention is paramount and prevailing with respect to the recognition as valid by other States of Certificates of Airworthiness, certificates of competency and licences issued by a contracting State. A standard for a similar recognition as valid of the AOC issued by another State was introduced in 2006 in Annex 6 (Operation of Aircraft). Standardizing certificates and licences and the recognition as valid is of utmost importance, especially in a time when civil air transport is evolving towards an increasingly complex regime with global properties where airlines are using aircraft registered abroad, in which flight crew are holding foreign licences and where (lease) aircraft are operated for long periods of time far from the State of Registry or the State of the Operator.

More complicated but highly topical is the situation in which operators are using company codes of several different countries or mergers of operators resulting in an air carrier with various operating bases located in different countries, let alone the outsourcing of a variety of activities related to air operations, subject to certification, licensing and recognition.

Although the Chicago Convention allocates certain functions to the State of Registry which that State is entitled to discharge, or obliged to discharge, where appropriate, the ICAO Assembly recognized, in Resolution A23-13, that the State of Registry may be unable to fulfil its responsibilities adequately in instances where aircraft are dry-leased, chartered or interchanged, in particular without crew or foreign crew members, by an operator of another State and that the Chicago Convention may not adequately specify the rights and obligations of the State of the Operator in such instances until such time as Article 83 bis of the Chicago Convention enters into force.[12]

If in the above-mentioned instances the State of Registry finds itself unable to adequately discharge the functions allocated to it by the Chicago Convention, it will delegate to the State of the Operator, subject to acceptance by the latter State, those functions of the State of Registry that can more adequately be discharged by the State of the Operator. This procedure will remain particularly relevant to contracting States that do not have treaty relations under Article 83 bis.

An aviation industry sector in which that particular diversity is less present and more manageable is the aerospace industry. The focus of the major aerospace facilities capable of building commercial jets is located in Europe, North America, Brazil, and some emerging markets such as the People's Republic of China.

Worldwide there are only a few countries able to design and produce heavy aircraft. To be able to built an aircraft there shall be in the first place an approved design consisting of such drawings, specifications, test reports, analyses and documentary evidence as are necessary to define the design of that aircraft and to demonstrate compliance with the design aspects of the appropriate airworthiness requirements. The preliminary Type Certification process consists of a Type Certification Basis to satisfy the regulator with respect to compliance with applicable rules and requirements, a Type Certification Board, a group identifying and evaluating the technical aspects on the basis of a proposed certification plan prior to a conditional or final approval of the design, and a Type Certification Program Plan in which the manufacturer indicates how to comply with relevant regulations.[13]

Normally, a request for a Type Certificate is formally submitted by the aircraft manufacturer when the aircraft is intended for serial production.[14]

The State of Design, most of the time being the State of the Manufacturer, shall issue a Type Certificate to define the design and to signify approval of the design of the aircraft type upon receipt of satisfactory evidence that the aircraft type is in compliance with the design aspects of the appropriate airworthiness requirements.[15]

---

12 20 June 1997. *See also* ICAO Doc 9318 *Protocol Relating to an Amendment to the Convention on International Civil Aviation (Article 83 bis).*

13 The FAA uses in the Type Certification Process the following sequence: Familiarization Meeting-Formal Application-Preliminary Type Certification Board-Certification Program Plan-Technical Meetings-Pre-Flight Type Certification Board-Type Inspection Authorization-Conformity Inspections and Certification Flight Tests-Aircraft Evaluation Group Determinations-Final Type Certification Board-Type Certificate-Post Certificate Activities.

14 A Type Certificate is a document issued by a contracting State to define the design of an aircraft type and to certify that this design meets the appropriate airworthiness requirements of that State. See also FAA Order 8110.4C, Type Certification.

15 ICAO Annex 8 Part II Chapter 1 Type Certification.

If a contracting State (or in the European Union: EASA) is convinced on the basis of satisfactory evidence that the designed and constructed aircraft complies with the design aspects of the appropriate airworthiness requirements, it shall issue a Certificate of Airworthiness.

However, the manufacturer, as one of the three major participants in aircraft certification and continued airworthiness process, will at any time be responsible for the original airworthiness of a newly launched (or a derivative-type design) aircraft type, aircraft engine or propeller.

The operator is primarily responsible for maintaining the expected level of airworthiness, even if the operator uses leased aircraft (from the manufacturer or a lease company) or aircraft subject to an interchange or charter agreement (with another operator) or relies on outside (outsourcing) contractors for maintenance.[16]

Under all circumstances, the national aviation authority (NAA) of a contracting State (or regarding the EU member States, EASA) will retain the responsibility for policing compliance with national aircraft certification, operating and maintenance regulations. Aircraft certification regulations also require the development of maintenance and quality control system requirements and operational limitations.

In the EU Basic Regulation (EC) No 216/2008, Article 4 requires EASA to take the responsibility for the design approval of products, parts and appliances designed, manufactured or used by persons/organizations under the regulatory oversight of EU member States, including the EEA States (Iceland, Liechtenstein and Norway) and Switzerland (EFTA member), except for products engaged in military, customs, police or similar services.[17]

Furthermore, EASA is responsible for the continuing airworthiness of aircraft and aeronautical products, parts and appliances, and on the approval of organizations and personnel involved in these tasks under Commission Regulation (EC) No 2042/2003 and its Annexes.[18]

In the European Union, not unlike the rest of the world, it is considered that products like an aircraft shall have a Type Certificate. The Type Certificate and certification of changes to that Type Certificate, including supplemental Type Certificates, shall be issued when the applicant has shown that the product (aircraft) complies with a type-certification

16 Committee on Aircraft Certification Safety Management, Aeronautical and Space Engineering Board, Commission on Engineering and Technical Systems, National Research Council, *Improving the Continued Airworthiness of Civil Aircraft. A Strategy for the FAA's Aircraft Certification Service*, National Academy Press, Washington, DC 1998, at 12.

17 *See also* for EASA responsibility in this respect: Article 4 (with exceptions in Annex II) Regulation (EC) No 1592/2002 of the European Parliament and of the Council of 15 July 2002 on common rules in the field of civil aviation and establishing a European Aviation Safety Agency, *Official Journal of the European Union L* 240/1, 7 September 2002.

18 *See also* EASA Part-M Subpart G continuing airworthiness management organization (CAMO).

basis as specified in Article 20 of the Basic Regulation, established to ensure compliance with the essential requirements referred to in paragraph I, and when it has no feature or characteristic making it unsafe for operation. The Type Certificate shall cover the product (aircraft), including all parts and appliances fitted thereon.[19]

In addition, according to Article 5.2c of the Basic Regulation, each aircraft shall be issued with an individual Certificate of Airworthiness when it is shown that it conforms to the type design approved in its Type Certificate and that relevant documentation, inspections and tests demonstrate that the aircraft is in condition for safe operation. This Certificate of Airworthiness shall remain valid as long as it is not suspended, revoked or terminated and as long as the aircraft is maintained in accordance with the essential requirements related to continuing airworthiness set out in point 1.d of Annex I (Essential requirements for airworthiness referred to in Article 5) and the measures referred to in paragraph 5 of the EU Basic Regulation.

Those essential requirements, applicable to aeronautical products (aircraft), parts and appliances, to persons and organizations involved in the operation of aircraft, as should be laid down by the European Community, are in line with standards and recommended practices set by the Chicago Convention and its Annexes. In this respect the Chicago Convention already provides for minimum standards to ensure the safety of civil aviation and environmental protection relating thereto. The essential requirements and rules adopted by the European Community for their implementation should ensure that member States fulfil the obligations created by the Chicago Convention, including their duties towards third countries.[20]

With respect to Acceptable Means of Compliance, technical interpretation of airworthiness codes and rules adopted by the European Commission to implement the Basic Regulation, and Guidance Material relating to Certification and Continuing Airworthiness in accordance with the provisions of the transparent, open rulemaking procedure for the development of opinions, certification specifications and guidance material (Article 52 Basic Regulation), the EASA shall establish consultation with the aviation authorities of the member States which perform an important role in assisting the EASA in rulemaking and ensuring compliance with international civil aviation standards in the field of certification and issuing guidance material on maintenance, operations and inspections, and other interested parties in the aviation industry, which are subject to rules drafted by the

19 Basic Regulation (EC) No 216/2008 Article 5.2a (Airworthiness).

20 *See also* Commission Regulation (EC) No 1702/2003 of 24 September 2003 laying down implementing rules for the airworthiness and environmental certification of aircraft and related products, parts and appliances, as well as for the certification of design and production organizations, *Official Journal of the European Union L* 243/6, 27 September 2003. *See also* Commission Regulation (EC) No 1194/2009 of 30 November 2009 amending Regulation (EC) No 1702/2003 laying down implementing rules for the airworthiness and environmental certification of aircraft and related products, parts and appliances as well as for certification of design and production organizations, *Official Journal of the European Union L* 321/5, 8 December 2009.

EASA, to create working groups in which each member State is entitled to designate an expert.[21]

EASA is also providing post-certification activities, like the approval of modification, replacement and or repairs of, aeronautical products and their components, as well as the issuing of mandatory Airworthiness Directives in order to be able to correct any potentially unsafe or hazardous situation.[22]

To ensure the continuing airworthiness of aircraft, including any component for installation thereto, which are registered in a member State or registered in a third country and used by an operator for which a member State ensures oversight of operations, Commission Regulation (EC) No 2042/2003 of 20 November 2003 on the continuing airworthiness of aircraft and aeronautical products, parts and appliances, and on the approval of organizations and personnel involved in these tasks, establishes common technical requirements and administrative procedures.[23]

In accordance with Annex I, Part-M.A. 301 of this Regulation, the continuing airworthiness of an aircraft shall be ensured by accomplishing any applicable Airworthiness Directives.

Essential requirements for continuing airworthiness shall be provided by instructions. These instructions must be established to ensure that the aircraft type certification airworthiness standard is maintained throughout the operational life of the aircraft and must be in the form of a manual, or manuals, as appropriate for the quantity of data to be provided.

The manuals must cover maintenance and repair instructions, troubleshooting and inspection procedures and servicing information, in a format that provides for a practical arrangement. The instructions for continuing airworthiness must contain airworthiness limitations that set forth each mandatory replacement time, inspection interval and related inspection procedure.[24]

For products, parts and appliances, for which the EASA only exercises the design responsibilities of the State of Registry, its policy is to endorse automatically the Airworthiness Directives issued by the State of Design, except if it itself issues a different Airworthiness Directive before the date of effectiveness of the State of Design Airworthiness Directive (Decision No 2/2003 of the Executive Director of the EASA of 14 October 2003 on the implementation of Airworthiness Directives for products, parts and appliances designed in third (non-EU) countries).

---

21 *See also* Commission Regulation (EC) No 2042/2003 Annex I Part M, Subpart C – Continuing Airworthiness, Annex II Part 145 – Approved Maintenance Organizations, Annex III Part 66 – Certifying staff-Maintenance, Annex IV Part 147 – Training organizations requirements.

22 Airworthiness Directives (AD) are issued by EASA, acting in accordance with Regulation (EC) No 216/2008 on behalf of the European Community, its member States and of the European third countries that participate in the activities of EASA under Article 66 of that Regulation.

23 Commission Regulation (EC) No 2042/2003 is amended by Commission Regulation (EC) No 707/2006, Commission Regulation (EC) No 376/2007, Commission Regulation (EC) No 1056/2008 and Commission Regulation (EC) No 127/2010 of 5 February 2010.

24 *See* Annex I of Regulation (EC) No 216/2008.

3.2 Licensing

> Article 16 Chicago Convention:
> The appropriate authorities of each of the contracting States shall have the right, without unreasonable delay, to search aircraft of the other contracting States on landing or departure, and to inspect the certificates and other documents prescribed by this Convention.

ICAO Annex 1 contains Standards and Recommended Practices adopted by the International Civil Aviation Organization as the minimum standards for personnel licensing. SARPs for personnel licensing were first adopted by the Council on 14 April 1948 pursuant to the provisions of Article 37 of the Chicago Convention and designated as Annex 1 to the Chicago Convention. They became effective on 15 September 1948. The Annex is applicable to all applicants for and, on renewal, to all holders of the licences and ratings specified herein.

The attention of contracting States is drawn to the obligation imposed by Article 38 of the Chicago Convention by which contracting States are required to notify the Organization of any differences between their national regulations and practices and the international standards contained in Annex 1 and any amendments thereto.

When a contracting State renders valid a licence, in particular a flight crew licence issued by another contracting State, as an alternative to the issuance of its own licence, it shall establish validity by suitable authorization to be carried with the former licence accepting it as the equivalent of the latter.[25]

A general rule is that the Licensing Authority shall confirm the validity of the other contracting State's licence before issuing the authorization in case an authorization for use in commercial air transport operations is to be issued under the rules of Annex 1.[26]

Another notable issue directly related to licensing and aviation safety is the international standard aviation language for communication. Especially in global air transport, with air routes within the airspace of so many countries with different languages, the standard English language proficiency associated with licences, as required by ICAO (as of 5 March 2008), has been adopted as being extremely important in a commutative sense.

ICAO, while refusing to extend its official March 2011 deadline for compliance with English proficiency requirements for airline pilots and air traffic controllers, is nevertheless insisting on a flexible approach towards a number of contracting States' civil aviation

25 Article 29(c) of the Chicago Convention requires that the flight crew members carry their appropriate licences on board every aircraft engaged in international air navigation.

26 ICAO Annex 1 Chapter 1.2 General rules concerning licences.

authorities that have yet to comply with these requirements established by a vote of the 36th Session of the 2007 ICAO Assembly. One of the requirements specifies: "that English shall be available on request at all stations on the ground serving designated airports and routes used by international air services". In October 2010, at its 37th Session, the Assembly, "recognizing that although the contracting States had made substantial efforts to comply with the requirements, some had encountered difficulty and wanted extra time. In response the Assembly again urged States to have their pilots and controllers use ICAO standardized phraseology in their communications".[27]

A majority of States expected to be compliant by 5 March 2011, although China had even recommended to extend the 2011 compliance deadline (the original deadline was scheduled for March 2008) or to adopt other transition measures to help ease the effort to ensure English proficiency. "China has consistently made unremitting efforts to implement ICAO requirements for English language used for radiotelephony communications. The majority of China's pilots engaged in international operations have met ICAO language requirements. However, due to the fact that a number of pilots are aging and their basic English language knowledge is limited, it still foresees some difficulties in the improvement of their language proficiencies within a short period of time", according to the Flight Safety Foundation AeroSafety World publication concerning ICAO standard aviation radiotelephony communication phraseology and proficiency requirements.

However, for aeronautical communication within the national borders some countries still prefer to continue using the native language as standard for domestic air transport, which could actually cause confusion and misunderstanding with regard to internationally operating foreign flight crews using English.

Aviation personnel, like aircrew who are required to use the radio telephone on board aircraft or like air traffic controllers and aeronautical ground station operators, shall demonstrate the ability to speak and understand the standard aviation (English) language used for radiotelephony communications to the level specified in the language proficiency requirements in Appendix 1 of Annex 1 to the Chicago Convention.[28]

27 L. Werfelman, *Speak Up*, AeroSafety World Flight Safety Foundation, December 2010-January 2011, at 41-42.

28 Avianca Flight 52, a Boeing 707-321B, crashed on 25 January 1990 in Cove Neck, Long Island, during the approach phase to New York JFK international airport, after running out of fuel. The final report of the NTSB on the accident determined the cause as pilot error due to the crew never declaring a low fuel emergency to ATC. Apparently the crew asked for a priority landing which, because of language differences between English and Spanish, can be interpreted as an emergency to Spanish-speaking pilots but not to English-speaking air traffic controllers. This may have caused confusion amongst the pilots when ATC confirmed their priority status.

With the extension of EASA's responsibilities to rulemaking in the fields of air operation and pilot licensing as well as the oversight of Third-Country Operators (TCO), EASA enhances its position as the European point of reference in aviation safety.[29]

EASA published on 27 August 2010 a proposal to the European Commission for a harmonized regulation on Flight Crew Licensing (Part-FCL) in the form of an opinion (No 04/2010), initially open to public consultation from June 2008 till February 2009 to comment on the subject, collected and published in April 2010 in the Comment Response Document (CRD). This new regulation, which will be adopted by the European Commission and enter into force as Community law by April 2012, will ensure that the same pilot licensing requirements and related high safety levels apply in all member States.[30]

3.3 Air Operator Certificate

A well-established method by which the State of the Operator may exercise the necessary control of its air operators is through the issuance of an Air Operator Certificate without which it is illegal to operate a commercial air transport service. In other words, an operator cannot engage in commercial air transport unless he holds a valid Air Operator Certificate issued by the State of the Operator according to Article 4.2.1 of Chapter 4 (Flight Operations) of Annex 6 to the Chicago Convention.

A certificate once issued should be subject to revocation if the operator subsequently fails or is unable to meet the conditions specified therein. Therefore, a State needs to establish a system for both the initial issuance of an AOC and the subsequent, continuing inspection and surveillance of operations. Such a system should be firmly based upon appropriate aviation law.

It is clear that the State of the Operator has major obligations to ensure air transport safety in general, and more specifically the safe transport of passengers and cargo on aircraft operated by the State's air carriers.

An issued AOC document shall at least contain the following data: the operator's location, date of issue and period of validity, description of types of operation authorized,

29 Third Country Operator means: any natural person residing in a third country or a legal person whose principal place of business is in a third country.

30 Community laws may take the following forms:
- *Regulations*, which are directly applicable and binding as part of the national law in all EU member States, and no action is necessary to transpose them into the national legislation;
- *Directives*, which bind member States to the objectives to be achieved within a certain time frame while leaving the national authorities the choice of form and means to be applied in implementing the Directives within their national legislation;
- *Decisions*, which are binding on those parties to whom they are addressed to any or all member States, to enterprises or to individuals, and;
- *Recommendations and Opinions*, which are not binding.

the types of aircraft authorized for use and authorized (international) areas of operation or routes. Prior to operating into another State, the operator from a particular State must submit detailed information to other States and sometimes obtain operations specifications from those States. An operator obviously has many responsibilities to ensure both safe operations and the safety and protection of passengers and cargo (ICAO Annex 6 Attachment F).

Before a flight is commenced, the operator is to use all reasonable means either through official published information or readily obtainable specifications and other data to ascertain that the anticipated flight is safe. An operator shall ensure that all facilities for the safe operation of a flight and the protection of passengers are adequate for the type of operation to be taken, that any inadequacy is reported without delay to the designated authority responsible, and that aerodromes are available throughout their published opening hours irrespective of weather conditions.

The State of the Operator regulations and rules for the operational certification of operators and the conduct of subsequent commercial air transport operations should be in conformity with the Annexes to the Chicago Convention and have sufficient detail to ensure that compliance will result in the desired level of safety. Article 16 of the Chicago Convention enables States to exercise their surveillance responsibilities regarding aviation operations conducted in their territory by foreign operators, using surveillance methods and inspections, without additional certification. With respect to the operator's AOC and related operations, continuing surveillance by the State of the Operator is inherent in the system of certification and is an essential part of the State's responsibility to ensure that the required standards of operations are maintained in order to provide a safe and reliable commercial air transportation service to the public. Adequate authority for certification and continuing surveillance of an Air Operator Certificate holder's operation should be contained in the provisions of the basic aviation law of the State.

The essential scope of Annex 6 (Operation of Aircraft) to the Chicago Convention is that the operation of aircraft engaged in international air transport must be as standardized as possible to ensure the highest levels of safety and efficiency.

The purpose of Annex 6 is to contribute to the safety of international air navigation by providing (constantly reviewed) criteria for safe operating practices, and to contribute to the efficiency and regularity of international air navigation by encouraging ICAO's contracting States to facilitate the passage over their territories (through their airspaces) of commercial aircraft belonging to other countries that operate in conformity with these criteria.

Annex 6 deals specifically with "Operations of aircraft". It reflects in detail the Air Operator Certificate requirements and the inspection process. It also deals with recurrency requirements of the AOC. Moreover, this Annex constantly refers to most, if not all (18 Annexes), other Annexes to ensure that the air operator will engage in and maintain safe operating practices. If, for example, during the required initial and periodic

AOC examinations it is found that the State Licensing practices are not in line with Annex 1 (Personnel Licensing) while all other aspects of the air carrier are found to be satisfactory, air safety is still at risk. This fact is also true for all other Annexes. Annex 6 is the only Annex that looks at aviation safety in its totality. Therefore, flight safety experts should not look solely at the performance of air operators, but their probe should equally focus on the whole spectrum in which the air operator actually functions.[31]

ICAO Doc 8335 also outlines the duties and responsibilities of both the State of the Operator and the air operator and recognizes their independence in maintaining an acceptable standard of operations. The organization, administration and procedures required for certification, inspection and surveillance of operators are outlined, including the establishment of an operations inspection organization, within the respective civil aviation authority. The development of a comprehensive Air Carrier/Air Operator (Operations) Manual is also stressed because of its importance in the certification process and day-to-day operations.

While references are included in the Operations Manual to financial matters to be considered by the contracting State authority prior to taking a decision on the issuance of an Air Operator Certificate, this Manual concentrates primarily upon the technical and air safety requirements to be met before issuance, and during the conduct of the operations authorized by the certificate. The system and procedures described are based on proven practices and, while directed at a wide range of international commercial air transport operations, are also deemed to be applicable to a wide range of national operations.

The Air Operator Certificate shall authorize the operator to conduct commercial air transport operations in accordance with specified authorizations, conditions and limitations. Contracting States shall recognize as valid an Air Operator Certificate issued by another contracting State, provided that the requirements under which the certificate was issued are at least equal to the applicable standards specified in Annex 6.

The issuing process is dependent upon the operator's ability to demonstrate an adequate organization, method of control and supervision of flight operations, training programmes as well as ground handling and maintenance arrangements consistent with the nature and extent of the operations specified. The State of the Operator, in addition to assessing the operator's ability and competence, should guide the operator in regulatory, organizational and procedural matters. The State of the Operator should be satisfied concerning the operator's eligibility for operational certification. This includes the ability and competence to conduct safe and efficient operations and proof of compliance with applicable regulations.

---

31 ICAO has issued guidance material to supplement their Annexes. Important in this respect is Doc 8335 AN/879, *Manual of Procedures for Operations Inspection, Certification and Continued Surveillance*, and Doc 9376, *Preparation of an Operations Manual.*

An operator must not only comply with Annex 6 but also with the provisions laid down in Annex 1, 8 and 18. In addition, the operator has to adhere to its AOC and associated Operational Specifications and if applicable to the authorizations to conduct specific operations issued by the State of Registry or the State of the Operator, and the authorization issued by EASA or the declaration it has made to EASA. Consequently, an operator shall not conduct operations when it is subject to a total operating ban pursuant to Regulation (EC) No 2111/2005 (Community list) or with one or more aircraft subject to a partial operating ban pursuant to the same applicable Community Regulation.

In the European Union with the first scaling-up of the Basic Regulation, the EASA became responsible for the operation of third-country operators.[32]

Third-country aircraft, as well as their crew and their operations, must comply with applicable ICAO standards. Just in case there are no such standards, the essential requirements laid down in Annex I, III and IV of the Basic Regulation will be applicable, provided that they are not conflicting with the rights of third countries under international conventions.

The regulatory framework for achieving the objectives laid down in Article 2 is provided by Article 18 of the Basic Regulation.[33]

EASA shall, where appropriate, issue opinions to the European Commission and issue certification specifications, including airworthiness codes and acceptable means of compliance, as well as any guidance material for the application of the Basic Regulation and its implementing rules.[34]

In this rulemaking process there is room for consultation (of interested parties and national authorities) and a comment and review period. In this democratic system NPAs (Notices of Proposed Rulemaking) are part of the EASA rulemaking process. EASA has developed a Comment Response Tool (CRT) that will be used to automate the consultation process for NPAs and Comment Response Documents (CRD). The CRT allows users

---

32 The full scope of competence of EASA so far, as established by the Basic Regulation, encompasses the regulation of airworthiness (Article 5) and environmental protection (Article 6), pilot licensing (Article 7), aircraft operations (Article 8) and aircraft used by third-country operators into, within, or out of the Community (Article 9).

33 The objectives of Article 2 can be summarized as:
- To establish and maintain a high uniform level of civil aviation safety throughout all the member States;
- To ensure a high uniform level of environmental protection throughout all the member States;
- To facilitate the free movement of goods, persons, and services;
- To promote cost-efficiency in the regulatory and certification processes;
- To promote Community views regarding civil aviation safety standards and rules throughout the world.

34 An Opinion is a draft of legislation (including amendments of the Basic Regulation and its implementing rules) that is sent to the European Commission. It is further processed either by the European legislator (the Council of Ministers and the European Parliament) or the European Commission before legislation is enacted. Certification specifications and guidance material for the application of Community law do not constitute mandatory requirements, and are simply the EASA's technical interpretation of Community legislation.

to review NPAs and place their comments, and later on to view the CRDs and add their reactions before the publication of the final Decision or Opinion. The TCO (Third-Country Operator) NPA is divided into three separate documents: Explanatory note and Regulatory Impact Assessment (RIA), Draft opinion and decision Part-TCO and Draft opinion and decision Subpart AR.TCO of Part Authority Requirements (AR).

Third country-operators engaged in commercial operations shall be subject to an authorization process in which they demonstrate their capability and means of discharging the responsibilities associated with their privileges to EASA.

The privileges granted to an operator and the scope of operations shall be specified in the authorization. If these operators intend to operate an aircraft not holding a standard ICAO Certificate of Airworthiness or with crew not holding a standard ICAO licence, they will be subject to an authorization process carried out by EASA before they can operate into, within or out of the Community.

Third-country aircraft, more or less permanently based in the territory of member States, and or operated by third country-operators, will be subject to the same rules and the same Community operational specifications as EU registered aircraft.

3.4 Concluding Remarks

Contracting or signatory States shall conform in accordance with the Chicago Convention to accomplish safe international air navigation. These signatory States should collectively work to harmonize and standardize the use of airspace for safety, efficiency and regularity of air transport. Each signatory State has a civil aviation authority to oversee the following areas of civil aviation which are paramount with respect to safe global air traffic: personnel licensing (regulating the basic training and issuance of licenses and certificates), flight operations (carrying out safety oversight of civil aviation air operators), initial and continuous airworthiness (issuing Certificates of Registration and Certificates of Airworthiness to civil aircraft, and overseeing the safety of maintenance organizations), aerodromes (designing and constructing aerodrome facilities) and air traffic services (managing the traffic inside of a country's airspace).

A number of European countries, both EU member States and Chicago Convention signatory States, have to conform to Community legislation in the field of civil aviation in continuation of the Chicago Convention and its Annexes (regarding aviation safety especially Annexes 1, 6 and 8).

In 2003 EASA was established as the central European Union regulator, taking over responsibilities for legislating airworthiness and environmental regulation from the national civil aviation authorities. EASA, as the single, coordinating aviation authority throughout the European Union, will eventually take over the power from individual EU member States to regulate all aviation, other than that specifically excluded from the scope of EASA. In other words, the Community has given EASA specific regulatory authority through the enactment of its regulations through the European Commission, the Council and the Parliament, as well as executive tasks in the field of civil aviation safety. It has jurisdiction over new type certificates and other design-related airworthiness approvals for aircraft, engines, propellers and parts. In the European Community, EASA works with the national aviation authorities of the EU member States, but has taken over many of their functions in the interest of aviation standardization throughout the European Union. Through standardization inspections it is responsible for monitoring the application of the European Union aviation safety regulations by the national aviation authorities of the EU member States.

Issues within the European Union such as flight crew licensing, Certificates of Airworthiness, Air Operator Certificates and assessment of aircraft used by third-country air operators, affecting the safety of air operations, are subject to standardization and centralization to better control the fast growing aviation industry.

Regulation (EC) No 1592/2002 (previous Basic Regulation) gave responsibility to EASA, considered as the centrepiece of the European Union's strategy for aviation safety, on 28 September 2003 for the airworthiness and environmental certification of all aeronautical products, parts and appliances designed, manufactured, maintained or used by persons under the regulatory oversight of EU member States. Some years later Regulation (EC) No 216/2008 (Basic Regulation) provided the Agency with additional operational responsibilities to air operations (the power to determine corrective actions and disseminate information to react without undue delay to a problem affecting the safety of operations), flight crew licences (responsibility for the approval and oversight regarding pilot training organizations and aero-medical centres and the certification of flight simulation training devices), and third country-operators (competency for the authorization and oversight of commercial operators and third-country aircraft when they or their crews do not comply with ICAO standards).

4 Coherent Aspects of European Union Safety Assessment of Foreign Aircraft

4.1 Civil Aviation Safety Assessment in the Netherlands as a Paradigm of the European Approach

Until 1991 there were no major currently known issues related to high-risk carriers operating to airports located in the Netherlands. However, this condition changed dramatically when on 21 October 1991, a Boeing 707 cargo aircraft of Gas Air Cargo, a Nigerian air operator, landed at Schiphol international airport. Shortly after the aircraft came to a full stop at the apron, the Dutch Aviation Police Service was taking a closer look at the decrepit aircraft.

Inspectors found some holes in the cabin floor and oil leaks in the hydraulic system. They further noticed a badly fitting cargo door. It proved to be very hard to put the cargo door in a closed and locked position. This decrepitude, however, was temporarily remedied by the crew through the attachment of a piece of wire. The cargo pallets could not possibly be put in locked position, while the loose captain's flight deck seat was secured in place by its own safety belts. And these deficiencies were only the most easily identifiable – certainly enough clues to what ultimately could be called a flying wreck.

As a logical consequence, the aircraft was grounded under Article 4 of the Dutch Aviation Act.[1] After provisional repairs, the empty aircraft was finally released for departure. The fine for the foreign air operator was relatively low: a maximum of 3,300 USD.

The incident, broadly covered by the media, drew attention to a potential safety issue: how is it possible that third-country rickety aircraft could enter Dutch airspace and land at an airport located near Amsterdam, being a densely populated area, and then, not only regarding the Netherlands but the European Community, how to prevent this phenomenon in the future? At that time, if there was no apparent reason, inspections of aircraft of foreign operators either did not take place at all or were relatively scarce.[2]

There was no reason whatsoever to be concerned about the safety condition of third-country aircraft.

---

1 As of 18 June 1992, this rule has been transferred to Articles 3.8 and 3.17 of the revised Dutch Aviation Act.

2 B. Meijer, 'Geen extra controle op verdachte vliegtuigen', *Haarlems Dagblad* (Dutch newspaper), 14 May 1996.

The Minister of Transport, Public Works and Water Management[3] (hereinafter referred to as "the Minister of Transport") responding to questions from the Dutch Parliament on the subject of assessment of airworthiness of aircraft operating under a foreign flag (State of Registry), said: "The ICAO system generally works well, in this respect that the probability that aircraft with arguable airworthiness reputations will operate in Dutch airspace, is quite low. (…) There is no reason for action".[4]

Two years later, a study (an objective analysis), commissioned by the Minister of Transport, was carried out by the RAND Corporation on the subject of air safety at Schiphol international airport.[5]

Regarding quality control of air carrier operators, the report provided the following conclusion: "It is fairly well known (and expressed to us in interviews) that some air carriers and aircraft types are more risky than others. Some aircraft of certain Eastern Bloc countries, especially those maintained in these countries, are considered quite risky. Some aircraft, pilots and maintenance associated with smaller third-world countries are considered less safe than those of major western countries and are not believed to meet international standards. (…) During the safety review at Schiphol, several parties indicated that the airport should be justified in posing minimum quality standards on its customers, particularly because of the huge investments involved, and the airport's responsibility to the public, which is exposed to the increased risks. (…) However, there are important limitations in the control that can be exercised in this area".[6]

The ICAO system did not work as satisfactorily as supposed by the Minister of Transport. Swift action was required. The Dutch Government decided, given the international aspects that play a role, to focus attention on this problematic nature to the international organizations in the field of aviation (ICAO, ECAC, JAA, EC).[7]

Apparently, there was little progress in this area, because when a few years later (1996) the Dutch aviation authorities were questioned whether a Soviet Ilyushin Il-76TD of Kazakhstan Airlines, that was chained up for nine months at Maastricht international airport, was still in an airworthy condition, the Dutch Civil Aviation Department stated that the aircraft was not allowed to leave unless the Kazakhstan authorities would declare the aircraft in compliance with the safety requirements. "Based on international standards",

3 The Ministry of Transport, Public Works and Water Management and the Ministry of Housing, Spatial Planning and the Environment in the Netherlands have been merged in 2010 to form a new Ministry of Infrastructure and the Environment.

4 Parliamentary questions concerning 'risky carriers'. Letter of 27 November 1991, RLD/DB/L 91.01211.9.

5 The RAND Corporation is a non-profit institution that helps improve policy and decision-making through research and analysis.

6 Final Report Safety Study Group Action Plan Schiphol, at 6.

7 'Jorritsma (Minister of Transport) Wants Tighter Control on Poor Performing Pilots and Aircraft', *Staatscourant* (Dutch State Gazette) 1995, at 220.

the Dutch Civil Aviation Department (from 1 July 2000 after a major reorganization in part continued as the Dutch Aviation Authority[8] on 1 July 2005 being merged in the Transport and Water Management Inspectorate (*Inspectie Verkeer en Waterstaat* IVW) Aviation Division), did not want to submit the freighter to an inspection by its own inspectors.[9]

The Dutch aviation authorities, at that time, felt that each State had to ensure that its air operators had to be compliant with international aviation standards and regulations. Meanwhile, this position has been abandoned (since 1997).[10]

Nowadays, the Aviation Division of the Dutch Transport and Water Management Inspectorate, which again has been renamed Human Environment and Transport Inspectorate (*Inspectie Leefomgeving en Transport* ILT) as per 1 January 2012 (hereinafter referred to as "the Transport Inspectorate"), owing to a merger of government departments, ensures that aircraft comply with airworthiness and environmental protection standards. Aircraft, engines, propellers, equipment, parts etc., must be properly designed, manufactured and maintained, in order to be permitted to operate in the airspace. Compliance is achieved mainly through certification of products and organizations. Only approved organizations have privileges to undertake aviation activities. Certification is increasingly performed within the international context, in which EASA plays a major role.

The Transport Inspectorate seeks to ensure safe, sustainable aviation within and above the territory of the Netherlands with as few incidents and instances of environmental pollution as possible.

The Transport Inspectorate therefore supervises safety and monitors compliance with the relevant environmental standards – for example, for noise abatement – at airports and the Dutch airspace. Regulation occurs in the form of on-site inspections and the issue of licences. Apart from Schiphol international airport, the Netherlands has a further 235 airfields. The Netherlands Air Traffic Control is also regulated by the Transport Inspectorate.

4.2 Safety Assessment Inspections

In the Dutch Civil Aviation Safety Policy Note, the Minister of Transport describes new and additional measures to make aviation even safer than before. In 1997 the Second Chamber of Parliament intensively debated and ultimately ratified this Civil Aviation Safety Policy Note.

---

8 The Dutch Civil Aviation Department was split into three independent divisions: the Dutch Aviation Authority, the Aviation Enforcement Service, including the SAFA-team and the Aviation Safety Counter, and the Directorate Aviation Policy.

9 *Dagblad De Limburger* (Dutch newspaper) of 26 and 29 June 1996.

10 *Dagblad De Limburger* (Dutch newspaper) of 12 January 1997.

One of the emanating results was the establishment of a specific SAFA team: a ramp inspection team that will accurately survey foreign civil aircraft landing at civil airports as well as at military airbases.

However, the deployment of a SAFA team in the Netherlands was not unique, as SAFA programme-based ramp inspections by specialized teams, affiliated to the ECAC and in partnership with the JAA, were deployed in other European countries too.

The intention of these aircraft safety ramp inspections still is to better understand all facets of aviation safety, but especially to detect and to directly deal with unsafe situations. Data derived from these safety inspections will be stored in a computerized European database managed by EASA (until 2007 by the JAA). This system is able to promote proper information exchange between States concerning underperforming air operators.

The safety inspections on foreign aircraft performed by the SAFA teams are aimed at (non)compliance with the SARPs established by ICAO.

The references for these inspections are contained in the standards of the ICAO Annexes 1 (Personnel Licensing), 6 (Operations of Aircraft) and 8 (Airworthiness of Aircraft), expanded to include Annex 18 (The Safe Transport of Dangerous Goods by Air) as an interface, due to the fact that on-board dangerous goods are part of the SAFA inspection items list (Part D). At first glance, air cargo is not considered an immediate cause for a fatal accident. Closer examination of potential hazardous materials gives rise to extremely careful treatment and checking.

With respect to the SAFA ramp inspection items concerning the aircraft's cargo compartments and dangerous goods (Part D),[11] the following items will be checked:

D1 General condition of cargo compartments
- Structures, wall panels, wall sealing
- Fire detection and extinguishing system
- Blow-out panels
- 9G-net
- Containers
- Loading instructions/door instructions
- Damages

D2 Dangerous goods
- Special authorization for the transport of DG
- Notification to the Pilot in Command
- Segregation and accessibility

11 EASA SAFA Guidance Material, Annex to ED Decision 2008/001/S, issue 1 of the Executive Director of the European Aviation Safety Agency of 29 September 2008 on the Guidance Material for the Qualification of SAFA Inspectors, at 26.

- Packaging and labelling (condition, leaks, damage, stowing)
- Limitations/Restrictions (cargo-only aircraft)

D3 Safety of cargo on board
- Loading instructions (placards, wall markings)
- Flight kit (secured)
- Pallets, nets, straps, containers (secured)
- Loading limitations (weight, size and height)

Ongoing (irresponsible) non-compliance with a number of standards of the ICAO Annexes could have significant unilateral consequences, while not responding to others could have far-reaching implications worldwide. Depending on the degree of the (repeatedly) failure of non-compliance with certain safety provisions of ICAO Annexes, fatal accidents could be the final outcome. Wrong application of, for example, Annex 18, although not exclusively (one can also think of unforgiving maintenance faults or fatal design flaws, aviation aspects corresponding to the areas covered by respectively ICAO Annexes 6 and 8 and associated guidance material and procedures), or not fully complying with its provisions, could be a breeding ground for a latent destructive condition, already present at the time of commencing a flight, eventually causing a serious aircraft incident or accident in just a totally different part of the world, affecting innocent people, just because of the neglect of rules. This lethal accumulation could most likely be avoided through targeted (cargo) aircraft safety inspections.

However, the SAFA safety inspections are, by their nature, rather limited in scope and profoundness. Their objective is to get a general impression on the safety standard of foreign aircraft, which in turn may reveal air carrier safety deficiencies because of (un) witting failure to comply with the adopted standards set out in the relevant ICAO Annexes. The extent to which air carriers are underperforming is often a reflection of poor governmental safety oversight. EU member States, along with ICAO and other specialized organizations, provide support and consultation to third country civil aviation authorities and other organizations in the areas of safety oversight to assist in fulfilling their safety oversight obligations, which can have far-reaching economic and political consequences for the country, the aviation authorities, and the air carriers.[12]

Strict governmental safety oversight will force air carriers to adhere to agreed safety standards to mitigate or rather eliminate the introduction of deadly errors.

12 The policy of the Netherlands is aimed at assisting underperforming countries regarding the restructuring of organizations of civil aviation authorities and the aviation industry, if required and on request. In this respect the Dutch Civil Aviation Department has started in 1997 in Africa a comprehensive programme, among others, regarding flight operations and Air Traffic Control Management. The focus is to cooperate with other western (assisting) countries.

4.3 SAFETY INSPECTION IN PRACTICE

The actual ramp inspection on foreign aircraft consists of three phases. In the planning phase it will be determined which operators are eligible for a survey by the SAFA team. Information, that leads to an aircraft subject to an inspection, could originate from various sources: from domestic or foreign airports, from other European aviation authorities, derived from the centralized SAFA-databank, and last but not least from concerned or complaining citizens who have contacted the Aviation Safety Counter.[13]

Apart from random inspections, the team determines, based on this information, which aircraft are to be inspected.

Following this determination, the execution phase will start. Three inspectors will initiate an inspection using a predetermined checklist, a procedure common to all participating States. Checks may include licences of the pilots, procedures and manuals that should be carried in the cockpit, compliance with these procedures by flight and cabin crew, presence and condition of mandatory safety equipment in the cockpit and cabin, cargo carried on board the aircraft, including the transport of dangerous goods, and the technical condition of the aircraft.

Directive 2004/36/EC Annex II includes a more detailed list concerning document and equipment items to be verified during ramp inspections in all member States:

I. The ramp inspection should cover all or a selection of the following aspects, according to the time available.
    1. Check for presence and validity of the documents necessary for international flights such as registration certificate, log book, certificate of airworthiness, crew licences, radio licence, list of passengers and freight.
    2. Check that flight crew composition and qualifications comply with the requirements of Annexes 1 and 6 to the Chicago Convention (ICAO Annexes).
    3. Check of operational documentation (flight data, operational flight plan, technical log), and of flight preparation necessary to show that the flight is prepared in accordance with ICAO Annex 6.
    4. Check for presence and status of items required for international navigation in accordance with ICAO Annex 6.
        - Air Operator Certificate
        - Noise and emission certificate
        - Operation Manual (including Minimum Equipment List) and Flight Manual
        - Safety equipment
        - Security cabin equipment

13 Citizens can also exert influence on the ad hoc checking of aircraft. Citizens who have observed unsafe or suspicious situations with certain aircraft or air operators can report to the Aviation Safety Counter. This office, which aims to increase accessibility for citizens and to activate the aviation safety information supply, as part of the Inspectorate, may pass such a report to the SAFA team for possible further action.

- Equipment necessary for the specific flight including radio communication and radio navigation equipment
- Flight data recorders

5. Check that the condition of the aircraft and its equipment (including damage and repairs) ensures continuous compliance with ICAO Annex 8 standards.

II. An inspection report must be drawn up after the ramp inspection and must include standard general information described above and a list of items checked, together with an indication of any deficiencies found for each of these items and/or any specific remark that may be necessary.

The SAFA team will accurately observe the exterior of the aircraft regarding obvious deficiencies, apparent corrosion, hydraulic oil and fuel leakage, general engine condition, windshield delamination, and tyres and brakes wear. This kind of inspection will take about 20 minutes while the aircraft is parked at the apron being serviced for the next flight. The short period of time needed for inspection will cause the least possible delay, if any.[14] Dependent on the seriousness of the findings, an operator will be inspected more frequently.[15]

4.4 Deviations

For each item to be inspected, four categories of possible deviations from the ICAO standard were initially defined, depending on its impact on flight safety: Category 0 means that no discrepancy is found; category 1 means that minor deviations were detected, such as a first-aid kit/emergency medical kit that is not located in the right place; category 2 means a serious shortcoming such as incomplete safety instructions, and category 3, which is considered a serious deviation such as blocked or obstructed emergency exits, an overload or unbalance situation, or technical defects.[16]

According to a recent report of the European Commission on the subject of categories regarding the severity of inspection findings, there are three designated categories: a category 1 finding is called a minor finding, a category 2 finding is a significant finding and a category 3 finding a major finding. These terms relate to the level of deviation from the relevant ICAO standard.[17]

14 With respect to SAFA inspections in the Netherlands: Information Magazine of the Ministry of Transport, Public Works and Water Management, No 4, April 1998.

15 Annual Report of the Dutch Transport and Water Management Inspectorate 2001, at 39.

16 Other shortcomings may exist in flight instruments, flight preparation, fuel calculations, weight (mass) and balance, (over)loading, licences and mandatory documents, various flight safety items in cockpit and cabin (lighting, fire extinguishers and oxygen masks), dangerous goods, and the general condition of the aircraft and its components.

17 Report from the Commission on the European Community SAFA Programme (Aggregated information-Report-30 April to 31 December 2006), 2008/C 42/01 of 15 February 2008, 4.2 Inspection findings and their categories.

The prime purpose of categorizing the findings is to classify the compliance with a standard and the severity of non-compliance with this standard. If the findings indicate that the safety of the inspected aircraft and its occupants is impaired, corrective actions will be required. It is standard practice that the captain of the inspected aircraft is debriefed about the findings, especially in case of a minor deviation for (less complicated) immediate corrective actions. In addition, categories 2 and 3 findings are communicated to the responsible civil aviation authority of the State of Registry and the home base of the operator (State of the Operator) with the request to take appropriate action because these findings require the highest attention when it comes to the need for rectification.

The Dutch Transport Inspectorate could impose corrective actions to prevent recurrence. Usually this will lead to departure delay until corrective actions are carried out in accordance with the requirements. The most drastic action the Inspectorate can take is to revoke the entry permit. This means that the particular aircraft is no longer allowed to land at Dutch airports or fly within Dutch airspace. Despite the sovereign nature, these measures have common similar effects in other EU member States, due to European harmonizing regulations.

The Transport Inspectorate has already used such an inexorable ban several times. In the spring of 2005 the entry permits of Phuket Air from Thailand and Turkish Onur Air were revoked, as was earlier the case with the Egyptian operator Flash Air. Fourteen days later Onur Air was allowed to resume activities in the Netherlands after setting up a plan for improvements, following consultation with the Transport Inspectorate.[18] In April 2004 the Transport Inspectorate suspended the entry permit of the Egyptian operator Luxor Air. An inspection revealed that Luxor Air did not meet certain safety standards such as the correct calculation of the aircraft's mass and balance, loading and the use of approved manuals. Luxor Air was not permitted into Dutch airspace until it demonstrated that it could meet the required standards.

In 2001 a total of 100 SAFA inspections were carried out in the Netherlands with 46 aircraft showing minor deviations. 15% of these inspections appeared to have one or more deviations from the ICAO standards with a major impact on flight safety (category 3), compared with 16.5% in 2000 (121 inspections). In recent years the number of inspections has increased from around 154 in 2003 with 39 category 3 findings to a total of 257 in 2007. In 2006 there were 93 category 3 findings (2005: 121) on a total of 266 SAFA inspections (2005: 243). The category 3 percentage remains around 15%, except for 2005 when the focus on suspected foreign aircraft was intensified, resulting in 243 inspections and 121 category 3 findings (Phuket Air, Onur Air and Fly Air). The number of findings

18 Germany, France and Switzerland also decided to impose a temporary ban.

per SAFA inspection on Onur Air in 2006 was remarkably well under the average of all inspected foreign operators in that year. Inspections are carried out mainly at Schiphol international airport, but increasingly at such regional international airports as Eindhoven airport, Maastricht Aachen airport and Rotterdam The Hague airport.[19]

In 2008 257 ramp inspections on foreign aircraft were carried out at Dutch airports. 13% of the inspections were in conformity with all requirements, while 34% needed corrective actions in order to be able and permitted to depart.[20] In case of serious deviations the Dutch SAFA team has the authority to take any appropriate action to prevent an aircraft from departing without being properly repaired.

4.5 Legal Aspects

The regulations based on the primacy of the authority of the State of Registry are internationally adopted. However, this does not mean that States will not be allowed to interfere in imposing requirements regarding foreign aircraft entering Dutch airspace. In bilateral civil air transport agreements, which the Netherlands has concluded with third countries, a provision is included stating that air operators must meet ICAO standards and, on request, have to demonstrate that:

> The Aeronautical Authorities[21] of a contracting party granting the operating authorization may require an airline designated by the other contracting party to satisfy them that it is qualified to fulfil the conditions prescribed under the laws and regulations normally and reasonably applied by such authorities to the operation of international air services.[22]

To emphasize this provision the following has been added:

> Each contracting party shall have to revoke an operating authorization of an airline designated by the other contracting party or to suspend the exercise of

19 Annual Report of the Dutch Transport and Water Management Inspectorate 2005, at 49. Regarding 2000-2004 *see* Annual Report 2004, at 55.

20 Annual Report of the Dutch Transport and Water Management Inspectorate 2008, at 32. In 2009 215 SAFA inspections were carried out in the Netherlands.

21 Aeronautical Authorities: in the case of the Russian Federation, the Federal Aviation Authority of Russia or any person or body authorized to perform any function presently exercised by the said authority and, in the case of the Kingdom of the Netherlands, the Minister of Transport, Public Works and Water Management or any person or body authorized to perform any function presently exercised by the said Minister.

22 *See also* the Air Transport Agreement between the Government of the Kingdom of the Netherlands and the Government of the Russian Federation, done in The Hague on 1 October 1997, Article 4.3.

> the rights specified in Article 3 of this Agreement or to impose such conditions as it may deem necessary on the exercise of these rights:
> a. In any case where it is not satisfied that substantial ownership and effective control of that airline are vested in the contracting party designating the airline and/or in its nationals; or
> b. In case of a failure by that airline to comply with the laws or the regulations in force of the contracting party granting these rights;
> c. In case the airline otherwise fails to operate in accordance with the conditions prescribed under the present Agreement.

In practice it means that before an operating authorization is granted by the Dutch authorities, the air carrier involved should submit its valid Air Operator Certificate, its insurance certificates and, if applicable, aircraft-type related noise certificates.

If in doubt, the Dutch authorities will contact the permit-granting Aeronautical Authorities concerned to make sure that the appropriate certificates are accurate and original and consequently in compliance with the minimum eligible requirements established by ICAO.

If there are still doubts, it is quite possible to conduct consultations with the relevant civil aviation authorities of the other contracting party under the bilateral Air Transport Agreement, yet to obtain this assurance and to resolve any questions. Finally, there is a possibility to revoke or suspend the rights of the air carriers concerned.[23]

4.6 Responsibilities under the Chicago Convention

The responsibility of a State is implicit in its acceptance of the international Standards and Recommended Practices for the safety of air navigation to which Article 37 (Adoption of international standards and procedures) of the Chicago Convention refers:

> Each contracting State undertakes to collaborate in securing the highest practicable degree of uniformity in regulations, standards, procedures, and organization in relation to aircraft, personnel, airways and auxiliary services in

23 The Netherlands has approximately 140 bilateral air transport agreements. However, these agreements, and other member State bilateral air transport agreements, will be corrected by new (horizontal) agreements concluded by the European Union (the European Union's exclusive external competence with regard to certain aspects of air services agreements concluded by member States with third countries). *See also* COM (2002) 649 final, Communication from the European Commission on the consequences of the Court judgments of 5 November 2002 for European air transport policy, 19 November 2002.

all matters in which such uniformity will facilitate and improve air navigation. To this end the International Civil Aviation Organization shall adopt and amend from time to time, as may be necessary, international standards and recommended practices and procedures dealing with:

a. Communications systems and air navigation aids, including ground marking;
b. Characteristics of airports and landing areas;
c. Rules of the air and air traffic control practices;
d. Licensing of operating and mechanical personnel;
e. Airworthiness of aircraft;
f. Registration and identification of aircraft;
g. Collection and exchange of meteorological information;
h. Log books;
i. Aeronautical maps and charts;
j. Customs and immigration procedures;
k. Aircraft in distress and investigation of accidents; and such other matters concerned with the safety, regularity, and efficiency of air navigation as may from time to time appear appropriate.

Although the methods for discharging its responsibility may vary, no particular method can, in any way, relieve the State of such responsibility.

The State's authority on flight safety activities is the respective Civil Aviation Department or Authority (CAA). The Operating Regulations should contain provisions for the Civil Aviation Authority to:

a. Determine the technical and financial capabilities of the operator to conduct the proposed operations with the ground services (*e.g.* aerodrome, navigation aids, air traffic services etc.) that can be made available;
b. Issue Air Operator Certificates, including the attachment of any Specific Operating Provisions he may determine to be necessary in the interest of safety of the particular operation;
c. Withdraw, revoke or vary operation authorized by the Air Operator Certificate should he think it necessary in the interest of safety to do so, subject to providing the operator with due notice and the opportunity for consultation and appeal;
d. Establish an operations inspection organization to assist in carrying out the functions and responsibilities of its office; and
e. Develop and amend Operating Directives consistent with the Operating Regulations.

In order to discharge its responsibility, the State must have aviation legislation that will provide for the development and promulgation of a code of operational regulations and practices consistent with its acceptance of the ICAO Annexes, and that will:

a. Represent a well-balanced allocation of responsibility between the CAA and the operators for the safety of operations;
b. Be capable of economic justification within the resources of the State;
c. Enable the CAA to maintain continuing regulation and supervision of the activities of the operator without unduly inhibiting his effective direction and control of his organization; and
d. Result in the cultivation and maintenance of harmonious relationships between the CAA and the operators.

4.7 Audit Management in General

Governments have the option of adopting provisions that will govern their role in the implementation of the operational regulations that may range from highly active to passive. In the active role, a close day-to-day interest would be taken in the direction and control of operations through an inspection organization. In the passive role, the State would intervene only to institute proceedings or investigative action when a violation of the regulations has occurred. In practice, neither of these extremes would be compatible with the division of responsibility between the State and the air operator.

A system of active supervision by the regulatory authority could be so rigorous as to amount to complete domination and dictation of the conduct of operations, leading to an undermining of the morale of operations personnel and a significant decline in operational safety. Such a system would, in any case, require the establishment of a large enforcement organization, which could hardly be justified. The State, in the passive role, could leave both the interpretation and the implementation of the regulations to the air operator, relying upon his technical competence to interpret them correctly, and encouraging compliance through threat of enforcement action.

This might place an unreasonable burden of responsibility on the air operator for interpreting and complying with the regulations. Also the State would not be in a position to assess the adherence of the air operator to these regulations other than by knowledge acquired fortuitously or in the course of an accident or incident investigation. Such a system would not enable the State to properly exercise the necessary preventive and corrective function, and consequently it could not adequately discharge its responsibility under the convention.

However, the aviation industry has changed quite substantially over the past few years, and, in fact, is still in transit. Deregulation and open-skies policies dramatically increased the number of operating air carriers worldwide. Rapidly expanding networks of relatively novice air carriers, emerging (inter)continental low-cost air carriers, intensified code-sharing, merging and air carrier alliances have far-reaching aviation safety implications, as the monitoring aviation authorities will be stretched to financial and human resources limits. The increasing complexity of both aircraft and the companies engaged in aviation has been accompanied by corresponding changes in the number of the requirements and the demands upon the regulators. The result is that the civil aviation authorities can no longer obtain an adequate picture of a whole operation without considerable, and progressive, enlargement of their resources. The scale of the increases that would be required is not affordable by the industry as a whole, nor would such an increase be the best response to the problem.

An alternative is to formalize the complementary nature of the duties of the civil aviation authority on the one hand and of those of the operator on the other. The respective civil aviation authority continues to be responsible for drawing up the regulations or the implementation of regulations imposed by Community institutions or global organizations, which form the base on which a safe operation is built, and also for overall surveillance. The operator remains responsible, among other things, by means of a safety management system, for the safety of the operation and for compliance with the requirements.

4.8 The Responsibility of the Air Carrier/Air Operator

The air carrier/air operator has a responsibility for the safe conduct of its operations and also a responsibility to comply with any regulations that the State may enact. These regulations, which may be regarded as the means by which the State wishes to implement the provisions of the ICAO Annexes, are not, however, in themselves sufficient to provide the air carrier/air operator with comprehensive and detailed instructions on which to base an operation.

A large measure of responsibility for the development of operating instructions considered necessary in their judgement for the safety, regularity and efficiency of their operations must therefore devolve upon the air carrier/air operator. These operating instructions must not conflict with the regulations of the State and of other States through which operations are conducted. The primary means used by the air operators to promulgate their operating instructions are the Operations and Maintenance Manuals. While only two documents are named, different manuals may be applicable in actual operations.

These manuals consist of operating instructions and procedures that are adjusted and fine-tuned to meet the air carrier's own specific management policies, however, well within prescribed safety and operating requirements. The Flight Operations Manual is a more detailed operations manual covering all relevant procedures regarding ground and flight operations.

Flight operations are easily the biggest item of air carrier expenditure, and they are directly responsible for flight safety action over the air carrier's flights. In the organizational structure of any Flight Operations Department there should be a flight safety section charged with internal first level monitoring of the air carrier's flight operations standards. Yet, flight safety should be considered the most cost-effective commodity in civil aviation when compared with the cost of an accident, although it is the most difficult to justify in terms of finance towards the air carrier management.

Training of aircrews is expensive, but the economics of training may be summed up by the air safety motto that an air carrier's best investment is to have well-trained crews. Such training should consist not only of actually piloting aircraft and its countless facets, but also of special training in mental attitudes towards flight safety, including Crew Resource Management (CRM), voluntary reporting of incidents and Human Factor's effect on aviation safety[24] – all as part of the air carrier's responsibility to assure flight operations safety.

4.9 OPERATIONAL CONTROL ELABORATED

According to ICAO Annex 6, operational control is the authority over the initiation, continuation or termination of a flight in the interest of aircraft safety and of the regularity and efficiency of the flight. It requires the operator to have a system in place to ensure monitoring of day-to-day flight safety activities and operations regularity.

However, it could also have a broader context. In the situation stated above, it is intended for the day-to-day operations of the air operator, although the State also should contribute substantially to create and maintain an overall safe aviation environment. Lack of up-to-date aviation rules and regulations, or poor aviation infrastructures and activities such as insufficiently equipped aerodromes, low-profile ATS and communications as well as unsatisfactory aviation security or negligent handling of dangerous goods will contribute significantly to an unsafe aviation environment.

Moreover, this could also severely jeopardize the regularity, the efficiency and the financial conditions of the air operator. Therefore, both the State and the air operator have responsibilities for the safe conduct of flight operations. In this context States shall establish a State Safety Programme (SSP) in order to achieve an acceptable level of safety in civil aviation, and as part of such a programme shall require that an operator implement a safety management system acceptable to the State of the Operator that, as a minimum:

24 *See also* Directive 2003/42/EC of the European Parliament and of the Council of 13 June 2003 on occurrence reporting in civil aviation, *Official Journal of the European Union L* 167/23, 4 July 2003, Article 9 Voluntary Reporting.

a. Identifies safety hazards;
b. Ensures the implementation of remedial action necessary to maintain agreed safety performance;
c. Provides for continuous monitoring and regular assessment of the safety performance; and
d. Aims at a continuous improvement of the overall performance of the safety management system.[25]

A safety management system shall clearly define lines of safety accountability throughout the operator's organization (flight operation, maintenance, etc.) including a direct accountability for safety on the part of senior management, well appropriate to the size and complexity of the operation.

The safety management system should as a minimum include:
a. A process or safety mechanism to identify actual and potential safety hazards and assess the associated risks;
b. A process to develop and implement remedial action necessary to maintain an acceptable level of safety; and
c. Provisions for continuous monitoring and regular assessment of the appropriateness and effectiveness of safety management activities.

The framework of a safety management system contemplates four components: safety policy and objectives, safety risk management, safety assurance and safety promotion. The implementation of such a framework shall be commensurate with the size of the organization and the complexity of the services provided. An almost identical format exhibits the framework of an SSP.

The implementation of an SSP is commensurate with the size and complexity of the State's aviation system, and may require coordination among multiple authorities responsible for individual elements of civil aviation functions in a particular State.

An air operator has many responsibilities to ensure both safe operations and the safety and protection of passengers. Specifically, an operator shall ensure that all facilities for the safe operation of a flight or series of flights and the protection of passengers are adequate for the type of operation to be undertaken, that inadequacy is reported to the authority responsible without delay, and that aerodromes are available throughout their published opening hours irrespective of weather conditions.

The regulations and rules set out by the State of Registry, or in case of inability to fulfil its responsibilities in certain instances, set out by the State of the Operator, should provide a framework of positive control and guidance but also allow the operator sufficient

25 ICAO Annex 6 Operation of Aircraft, Part I Commercial Air Transport Aeroplanes 3.3 (Amendment 33-B).

flexibility to develop and update instructions for the detailed guidance of personnel essential to the conduct of the operator's operations.

Furthermore, the State should guide the operator in regulatory, organizational and procedural matters in view and in addition to assessing the ability and competence of the operator to conduct safe and efficient operations and proof of compliance with applicable regulations. On the other hand, the State itself has to comply with internationally adopted minimum safety requirements and regulations in order to maintain proper safety oversight.

While there is no magic formula available that will guarantee full compliance by contracting States with ICAO safety oversight provisions, it is important that those obligations are properly documented in Annex 6 and that they have been laid down in a comprehensive and clear manner.

A second important step is to greatly enhance awareness of these obligations at higher levels in contracting States and prescribe alternative, practical means to meet those obligations.

When drawing attention to the safety mechanisms of an air operator, Annex 6 recommends (from 1 January 2002), among other things, that an operator of an aeroplane of a maximum certified take-off mass in excess of 20,000 kg should establish and maintain a flight data analysis programme (from 1 January 2005 in excess of 27,000 kg shall establish) as part of its safety management system (Annex 6 3.3.5 Recommendation) and that a flight data analysis system shall be non-punitive and contain adequate safeguards to protect the source(s) of the data, which could be extremely important and delicate in view of just culture aspects for those source(s). Especially effective safety data monitoring (collection, analysis and exchange) is one of the eight critical elements of the State's safety oversight functions, of which some designated ones are focused on the service provider's safety performance.[26]

The disclosure of confidential safety data, either derived from accidents, (voluntary) incident reporting or flight data management, should be protected by the State in such a way that broad access to these data shall be limited in the interest of dedicated front-line professionals. Protection of data, a State oversight function, must be an undeniably conscientious task performance.[27]

---

26 The term "service provider" refers to any organization providing aviation services. It includes approved training organizations that are exposed to safety risks during the provision of their services, aircraft operators, approved aviation maintenance organizations, organizations responsible for type design and/or maintenance of aircraft, air traffic services providers and certified aerodromes, as applicable (Annex 6 Operation of Aircraft Amendment 33-B).

27 *See also* Directive 2003/42/EC of the European Parliament and of the Council of 13 June 2003 on occurrence reporting in civil aviation.

4.10 Government Oversight of Air Operations

During the early years of aviation, commercial aviation, without the necessary experience, was a loosely regulated and organized activity characterized by underdeveloped technology, lack of proper infrastructure, limited governmental oversight and an insufficient understanding of the hazards underlying aviation operations. As a consequence, those days were depicted as potentially hazardous, which is reflected in numerous aircraft accidents.

The overriding priority of the early safety process was the prevention of accidents, and basic accident investigation was the principal means of prevention. Technological improvements, development of an appropriate infrastructure and an increasing drive to heavily regulate civil aviation, resulted in a rather safe transport system. The still pervasive notion that safety can be guaranteed as long as rules are followed and that deviation from rules necessarily leads to safety breakdowns, is straightforward but subject to limitations inherent to the increasing complexity of an operational system as open and dynamic as civil aviation.

Transition from a predominantly prescriptive regulatory environment to an integrated regulatory environment combining prescriptive and performance-based regulatory approaches, is fundamental and must start by clearly establishing the role of the State's safety oversight function within the SSP and their mutual relationship.

Fundamental to State safety oversight is availability of proper aviation rules and regulations. This element became implicit when the State signed the Chicago Convention. As is normal in any State, making rules and regulations is complex and time consuming. This is particularly so in the developing States. A national safety legislative framework and specific regulations, in compliance with implemented international and national standards, are a prerequisite for proper safety oversight and are certainly an issue that cannot be left to the air carrier/air operator. While there are basic rules and regulations necessary, the complexity of the aviation requirements of the State will, to a large extent, dictate the follow-on rules and regulations required to efficiently and effectively oversee safety.

The responsibility of a State is implicit in its acceptance of the ICAO SARPs for the safety of air navigation.[28] Given the main objectives of the safety oversight function, the common components of a national safety oversight system may be summarized as follows:

a. Monitoring of safety performance;
b. Verifying compliance with applicable safety regulatory requirements;
c. Safety regulatory auditing;
d. Oversight of new or changed systems, operations, products or procedures;
e. Publication of regulatory instructions or advisory material based on findings of oversight activities;
f. Generation and maintenance of safety oversight records.

28 *See* Article 37 of the Chicago Convention.

The management of aviation safety in a State includes the participation of the relevant civil aviation authority or its equivalent in specific activities related to the management of aviation safety in the State, and the establishment of the roles, responsibilities and relationships of organizations in the system.

4.11 SAFETY CLAUSES IN BILATERAL AGREEMENTS

In the context of negotiations between States regarding air traffic rights, stringent arrangements have been made since 1997 about safety requirements that must be met by aircraft flying between these States. In bilateral civil air transport agreements an additional safety clause will be included with more extensive safety requirements than before. Such a specific clause is an additional instrument to support the oversight of the relevant permit-granting aeronautical authorities of the contracting parties. This safety clause differs from the standard approach of earlier agreements, which all required that aircraft in question should meet minimum standards.[29]

According to the Minister of Transport, it does not mean that the absence of such an additional clause "would not provide an opportunity to the Dutch Government to inspect foreign aircraft after landing at Dutch airports and if needed to take appropriate action in case there are serious doubts concerning the safety status of these aircraft".[30] Such a statement, in fact, will not be confined to the Netherlands, but asserts itself in all EU member States.

Based on a government decision of 12 May 2005, the State Secretary of Transport has imposed a flight ban on Turkish Onur Air to enter Dutch airspace.

The imposition of a flight ban on Onur Air was motivated by the concern present within the Dutch Aviation Division of the Transport Inspectorate about the way in which Onur Air in its daily practice of flight operations was dealing with flight safety standards applicable under the international civil aviation regime, which also includes Onur Air as a regular air carrier. This concern was based on a long series of incidents and findings and, as a consequence, an emerging pattern of negligence and carelessness. The State Secretary of Transport has based its competency on Article 6 of the Resolution Non-Scheduled Air Transport of 2 May 1975: suspension of a route permit due to safety aspects.

Onur Air has objected to the competency of the State Secretary in the appeal stage because the State Secretary implicated the aspect of safety in the consideration which formed the basis for the intervention.

However, the State Secretary was of the opinion that the permit was revoked in the correct way. The motivation was as follows:

29 In particular, the standards in Annexes 1, 6 and 8 to the Chicago Convention.

30 Letter of the Dutch Minister of Transport to the chairman of the standing House committee of the Second Chamber of Parliament of 13 May 1997.

The Chicago Convention creates rights regarding unrestricted operation of aircraft engaged in international commercial air transport. However, these rights coincide with obligations, and moreover at the same time will not impede the responsibility and authority of contracting States to establish their own regulations with respect to entry and departure of aircraft in their territory and requirements which aircraft must meet while operating within their territory. One of the aspects where States preserve their own responsibility and authority is the safety of air traffic and aircraft, as far as its own territory is at issue. Multilateral and bilateral agreements concerning freedom rights do not restrict this authority but consider this authority as sovereign.[31]

Aircraft of foreign air operators engaged in international air navigation admitted to enter, be within or depart from the territory of another contracting State must, without exception, abide by the national regulations and laws of that State, as reflected in Article 11 of the Chicago Convention:

> Applicability of air regulations. Subject to the provisions of this Convention, the laws and regulations of a contracting State relating to the admission to or departure from its territory of aircraft engaged in international air navigation, or to the operation and navigation of such aircraft while within its territory, shall be applied to the aircraft of all contracting States without distinction as to nationality, and shall be complied with by such aircraft upon entering or departing from or while within the territory of that State.

Article 11 requires non-discriminatory application of law and regulations of a contracting State or a region consisting of a number of contracting States such as the European Union on international flights. Within the European Union non-compliance will eventually be penalized through inclusion on the Community list (EU blacklist).

4.12 EU Blacklist

If a foreign air operator exhibits a pattern of defects and incidents and the foreign authority responsible for oversight will not forestall this situation, the publication of a blacklist indicating designated unsafe operating air carriers will be a means to put compelling pressure on a blacklisted air operator with the aim of remedying shortcomings and subsequently of enhancing its operational safety level.

31 Apology on the subject of the dispute between the Dutch State Secretary of Transport and Onur Air of Turkey.

For this purpose the European Commission established, on 22 March 2006, a Community list of air operators which at that time were totally banned from flying to destinations within the European Union. This initially formatted blacklist of unsafe operating air carriers was based on Regulation (EC) No 2111/2005 of the European Parliament and of the Council of 14 December 2005 on the establishment of a Community list of air carriers subject to an operating ban within the Community and on informing air transport passengers of the identity of the operating air carrier, and repealing Article 9 (Imposition of a ban or conditions on operation) of Directive 2004/36/EC (SAFA Directive). It should be noted that the operating bans included in the Community list should apply throughout the territory of the member States to which the Treaties of the European Union apply. In Regulation (EC) No 2111/2005 legal obligations are defined, in addition to contract of carriage performance obligations, which, however, are governed by the laws of the member States applicable to contracts and by relevant Community law. The same applies for Commission Regulation (EC) No 473/2006 of 22 March 2006 laying down detailed implementing rules for the Community list of air carriers that are subject to an operating ban within the Community in respect of the procedures referred to in Chapter II of Regulation (EC) No 2111/2005.

The Community list should be based on common criteria for consideration of an operating ban for safety reasons drawn up at Community level. Before drafting and adopting this particular blacklist Regulation, a committee of aviation safety experts assessed all national flight groundings (operating bans).[32] The Community list almost immediately entered into force as of 22 March 2006. Each member State shall enforce, within its territory, the operating bans included in the Community list in respect of the air carriers that are the subject of those bans. Regulation (EC) No 2111/2005 establishes rules in the following parts as stated in Article 1 (Subject matter):

a. On the establishment and publication of a Community list, based on common criteria, of air carriers which, for safety reasons, are subject to an operating ban in the Community; and,
b. On informing air passengers of the identity of the air carrier operating the flights on which they travel

All European citizens must have access to this particular information.

Total or partial operating bans shall be imposed on an air carrier only when it is demonstrated (and verified) that the objective and transparent common criteria, based on the relevant safety standards, as stated in Regulation (EC) No 2111/2005, are not met. The operating ban affects, in particular, air carriers from Africa, Central-Asia and the Far East. Their names can be found on two blacklists, which replaced the previous national lists. Air passengers are better protected by the Community list. In addition, this list, contrary

32 Newsletter of the Dutch Ministry of Transport, Public Works and Water Management of 22 March 2006.

to previous diverse national lists, will provide similar protection throughout the European region.[33]

The European Commission, acting on its own initiative or at the request of a member State, shall decide to update the Community list as soon as this is required in the following cases:

a. To impose an operating ban on an air carrier and include this air carrier on the Community list, on the basis of the common criteria;
b. To remove an air carrier from the Community list, if the safety deficiency or deficiencies that gave rise to the inclusion of the air carrier on the Community list, have been remedied and, moreover, there is no other reason, on the basis of the common criteria, to maintain the air carrier on the Community list;
c. To modify the conditions of an operating ban imposed on an air carrier which is included on the Community list.[34]

This updating shall be in accordance with the procedure referred to in Article 15(3) of Regulation (EC) 2111/2005 and on the basis of the common criteria.[35]

At least every three months, the European Commission shall verify whether it is appropriate to update the Community list. On 19 April 2011 the Commission has updated the list of carriers banned from operating within the European Union for the 17th time, bringing the total number of banned air carriers to 269.[36]

The Dutch Ministry of Transport has actively contributed to the creation and adoption of the Community list. Air carriers, already banned by the Dutch aviation authorities, have been banned as from 22 March 2006 across the European Union.

Before the European Commission introduced the Community list, extensive discussions have taken place at the national level in all member States regarding the usefulness and necessity of a single, common blacklist. The Netherlands expressed itself against such a list.[37] On the other hand, the Netherlands supported the proposal of the European Commission to create a joint blacklist of air carriers that are not allowed to operate to, from and within the European Union.

Not everyone welcomes the blacklist introduced by the European Commission. According to the Dutch Forum for Sustainable Aviation (*Platform Duurzame Luchtvaart*

33 In 2005 the Turkish air carrier Onur Air was subject to a temporary operating ban in the Netherlands and France, while their aircraft with French and Dutch passengers continued daily operation to the Belgian international airport of Charleroi.

34 Article 4 (Updating of the Community list) of Regulation (EC) 2111/2005.

35 Articles 5 and 7 of Decision 1999/468/EC (Council Decision of 28 June 1999, laying down the procedures for the exercise of implementing powers conferred on the Commission) shall apply, having regard to the provisions of Article 8 thereof.

36 UK CAA Press Office, 21 April 2011.

37 Parliamentary questions and answers 2004-2005, No 2424 TK.

PDL),[38] fearing a false sense of safety, a blacklist will not provide any indication on the likelihood of an air transport passenger getting involved in any aircraft accident. An aviation accident is, apart from a possible preconceived lethal action, a concurrence of coincidental circumstances.

It is not inconceivable that an air carrier can be placed on a blacklist without ever having experienced an aircraft accident. A reverse situation may also arise: an air carrier can meet all operational safety requirements and still suffer from an aircraft accident. However, there might be little choice for air carriers not willing their potential passengers to consider flying with other carriers, especially for safety reasons. They will be bound to improve the level of flight safety. However, in marginal business conditions there will also be air carriers that are reluctant to report mishaps or are neglecting their focus on safety. Unfortunately, this trend seems to worsen in years of economic crisis.

It must be clear that no lessons can be learned from incidents and deficiencies that are not reported at all. On the contrary, they will ultimately have adverse effects.

The fact is that IATA opposes the issuing of a blacklist, because a blacklist would not immediately assist in improving aviation safety. IATA's view is rather a continuous improvement of quality and safety and, accordingly, it has developed the IATA Operational Safety Audit project (IOSA) and programme, where air carriers affiliated to IATA will be monitored for compliance with the criteria set by IATA.[39]

IOSA is intended to formulate and implement an internationally recognized evaluation system, by which the level of competence and reliability of an air carrier to deliver a safe operation and manage attendant risks may be established.[40]

Moreover, the US FAA has been issuing its own blacklist since 1 February 1997. This list concerns States, *casu quo*, aviation authorities, but has no direct relationship with foreign air carriers.

FAA's rationale is that if the relevant civil aviation authority service present in a State is not sufficiently and appropriately equipped, the safety level of the respective national air carrier will not satisfactorily be guaranteed either.[41]

4.13 Oversight with Respect to Foreign Aircraft Based in the Netherlands

Not all aircraft operated by Dutch air carriers are registered in the national civil aircraft register. At various airports in the Netherlands there are a number of aircraft with foreign

38 *See also* SANE (Sustainable Aviation Network Europe).
39 *See*: <www.iata.org/ps/services/iosa/index.htm>.
40 ICAO Assembly Working Paper A33-WP/83 EX/33, 9 August 2001.
41 Letter of the Dutch Minister of Transport to the Chairman of the Second Chamber of Parliament, Parliamentary Documents II, 24 804, No 3, 20 December 1996.

(American, even Russian) registration marks – in most cases, private or business aircraft, but to some extent commercial jetliners as well. Monitoring and oversight on maintenance and (continuous) airworthiness will be the responsibility of the State of Registry, according to international adopted agreements. Oversight on foreign registered aircraft, stationed for a longer period in the Netherlands, is generally imperfect or rather below standards, according to a conclusion of the Dutch Safety Board (since 1 February 2005 the legal successor of the Dutch Transport Safety Board) following two accidents in the Netherlands in 2001 and 2002 involving two Russian registered general aviation aircraft.[42]

The Dutch Safety Board limited its investigation to the extent of the general aviation sector only. Worldwide, the ICAO SARPs provide guidance in determining the aviation legislation at the national level. In general, the rules and regulations applicable to the operation of the aircraft are those that are in force in the State of Registry. Within Europe the Joint Aviation Authorities (JAA) contributed to the harmonization of legislation between the JAA member States to an important extent. This tendency is continued within the EASA (since 2003) by the EU member States that have joined the JAA. This has resulted in uniform JAA and/or EASA regulations, among other things, applicable to aircraft registered in a member State. At the time the two accidents occurred, JAA regulations were applicable. For the record it is emphasized that Russia is not a JAA member State.[43]

Annually, approximately fifty Dutch and foreign registered (though semi-permanently stationed in the Netherlands) aircraft are inspected at random. The inspection largely consists of checking the technical status of an aircraft and the validity of aircraft documents. The Dutch Safety Board, however, observed in 2005, by means of an oriented investigation, a gap in the enforcement of regulations regarding foreign registered aircraft originating from outside the effective area of the JAA and EASA but (semi-permanently) stationed at Dutch airports. The Safety Board investigation revealed that these aircraft were not certified in compliance with international civil aviation standards and that outside the Russian Federation oversight of Russian registered aircraft was totally absent, as a result of which the random inspections of the Transport Inspectorate had to secure flight safety in this respect.

However, the Safety Board clearly observed that such random inspections may have insufficient proficient depth to guarantee flight safety.

The Dutch Safety Board concluded that the State in which foreign registered aircraft (permanently) are stationed and operated should in fact have equal responsibilities as the State of Registry. That means in this case that the Dutch law and regulations should apply

42 Yakovlev Yak-52, registration RA 02080, on 9 March 2001 and Sukhoi SU-29, registration RA 01276, on 7 June 2002 (Dutch Safety Board reports 2001028 and 2002079).

43 In May 2002, after the sale of both aircraft, the British Civil Aviation Authorities indicated that the Russian authorities informed them of the fact that Russian registered aircraft starting with RA01, RA02 and RA44 may only be operated in Russia (Dutch Safety Board accident report, Conclusion 6).

to foreign registered aircraft as soon as they exceed a certain continuous stationing period in the Netherlands.[44]

Needless to mention, in fact, that, in accordance with Annex 13 of the Chicago Convention as well as EU Council Directive No 94/56/EC of 21 November 1994, establishing the fundamental principles governing the investigation of civil accidents and incidents, the (technical) investigation of the Dutch Safety Board is not intended to apportion blame or liability.

4.14 AVIASSIST

Another support programme for third countries in the field of aviation safety is AviAssist, a programme well known in the Netherlands and beyond.[45] As a regional affiliate of the Flight Safety Foundation, the independent AviAssist Foundation identifies threats to aviation safety, analyzes the problem issues and works on practical solutions to these cases.[46] The Foundation, being non-profit and thus reliant on support and subsidy from governments, corporations, individuals and foundations as well as fees for specific services, provides proactive safety support to civil aviation organizations in the 23 contracting States of the ICAO East and Southern Africa region in order to ensure the infrastructure to be at the (ICAO) minimum safety level. AviAssist is committed to work in coordination and in accordance with ICAO's Comprehensive Regional Implementation Plan for Aviation Safety in Africa (ACIP).[47]

The background of this support programme is not only to guarantee aviation safety in the Netherlands, but, as a matter of fact, to ensure a safe operation environment for Dutch aircraft with respect to foreign destinations in specific regions.

Statistics show that the probability of an aviation accident over the African continent is considerably larger than in the European Union. To reduce this risk the Ministry of Transport decided to devote special attention to the African continent in this respect by introducing the support programme called AviAssist.

If a Dutch air carrier intends to start operations to a State that hardly meets the minimum ICAO requirements, the Ministry of Transport will attempt to make improvements

44 *Staatscourant* (Dutch State Gazette) 2005, No 20, at 3.

45 The Aviassist Foundation is registered in Hoofddorp, the Netherlands. The Foundation's main office is based in the United Kingdom. This support programme originated the Note Safety Policy Civil Aviation prepared by the then Dutch Director Civil Aviation Inspectorate. The aim of this Note was to achieve general approval for this policy and to inform the Dutch Parliament and interested civilians regarding plans and ideas of the Transport Inspectorate to make civil aviation more safe.

46 The Flight Safety Foundation was formed in 1947 as an independent, impartial and non-profit international membership organization, to pursue the continuous improvement of global aviation safety. The FSF headquarters are in Alexandria, Virginia, United States of America.

47 *See also* AFI (Africa-Indian Ocean Plan) Planning and Implementation Regional Group Sixteenth Meeting (APIRG/16), Kigali, Rwanda 19-23 November 2007 (APIRG/16-WP/9 14 November 2007).

in that State by means of setting up a local project. Basically, the assumption must be made that all parties active in civil aviation have their own responsibility and that AviAssist (as a section of the Transport Inspectorate) will not take over that responsibility. According to the JAR-OPS, air carriers have their own responsibility to make sure that the destination to operate flights to will be safe enough. The air carriers themselves shall have to strike a balance between safety and economics. AviAssist will, among other things, come into action on the request of air carriers. These air carriers, invoking assistance, are mostly of the opinion that the minimum safety standards regarding a given destination are in jeopardy and that subsequent operations to that specific destination are not to be continued without proper improvements regarding aviation safety.

If a country meets the criteria determined by AviAssist to provide assistance, a number of steps must be followed. As far as the Netherlands is concerned, the most obvious criteria are: there must be some Dutch interest; it must fit within the policy framework of the Dutch Ministry of Transport and the purpose must be the enhancement of basic safety standards. The first step is to make an inventory of the problematic issues. The result will be a project plan (including planning, financing and objective), which will be materialized by means of successive steps aimed at improving practice. The starting point of all improvements is that the ICAO SARPs will be considered the basic standard format that all countries should meet.

A model of one of these projects is Tanzania: KLM has been operating for many years now to Tanzania, but the destination airports like Dar es Salaam and Kilimanjaro did not meet the required safety standards in the early years of operation (insufficient fire extinguishing capability, poor navigation aids and communication equipment, etc.). As a result of the project, new crash-tenders are purchased, navigation and communication equipment is renewed, maintenance contracts have been concluded and the staff is regularly trained. In general, the Dutch Ministry of Transport is trying to build a long-term relationship with a third-world country (in this case Tanzania), so that the result exhibits a sustainable nature.

4.15 International Airworthiness Requirements

Article 31 of the Chicago Convention determines that every aircraft engaged in international navigation shall be provided with a Certificate of Airworthiness, issued or rendered valid by the State in which it is registered. Article 33 further submits that such a Certificate of Airworthiness, issued or rendered valid by the contracting State in which the aircraft is registered, shall be recognized as valid by the other contracting States, provided that the requirements under which such a certificate was issued or rendered valid will be equal to or above the minimum standards that may be established by ICAO from time to time pursuant to the Convention.

The standards in Annex 8 (Airworthiness of aircraft) define the minimum level of airworthiness to be met by the contracting States. In addition, there are standards that specifically address the obligation of contracting States in a particular capacity, as for example the State of Design: The State of Design of an aircraft shall transmit any generally applicable information that it has found necessary for the continuing airworthiness of the aircraft and for the safe operation of the aircraft to every contracting State that has advised the State of Design that it has entered the aircraft concerned on its register (State of Registry) or to any contracting State upon request.[48]

Through Council Regulation (EEC) No 3922/91 of 16 December 1991 on the harmonization of technical requirements and administrative procedures in the field of civil aviation, the technical Joint Aviation Requirements of the JAA, applicable to certain types of aircraft, engines and propellers, or parts thereof, have been incorporated by implementation in the Dutch Aviation Act.[49] With the creation of EASA, JARs are gradually being amended as necessary and brought into European legislation so that they are binding on all EU member States.

The JAA requirements were well in line with ICAO standards, but much more stringent.[50]

In all phases of aircraft operations the rule will apply that, if a Certificate of Airworthiness has been issued once, the holder of an aircraft has the obligation to ensure the continuous airworthiness of that aircraft. In order to ensure continuous airworthiness, the holder must use approved maintenance companies that have qualified technicians. The national civil aviation authorities are responsible for monitoring and compliance. If there is any doubt regarding the airworthiness of an aircraft, for example in the event the aircraft has sustained damage, the State of Registry shall judge whether the damage is of a nature such that the aircraft is no longer airworthy, as defined by the appropriate airworthiness requirements.[51]

4.16 NATIONAL AIRWORTHINESS REQUIREMENTS

The Dutch aviation legislation with respect to airworthiness, largely based on the Joint Aviation Requirements, has been replaced by EU legislation.

48 Articles 4.3.1 and 4.3.2 of ICAO Annex 8 Airworthiness of Aircraft.

49 Prior to the creation of EASA, the Joint Aviation Authorities (JAA), as an organization, was responsible for publishing regulations, the Joint Aviation Requirements, governing the operations, maintenance, licensing and certification and design standards for aircraft. JARs had no legal status until they were ratified by national governments. Many European nations adopted all or part of JARs as the basis for their own national regulations.

50 On the basis of a decision of ECACs DGs in adopting the FUJA II Report it was decided to disband the JAA system per 30 June 2009.

51 ICAO Annex 8, 3.6.1. Damage to Aircraft.

Article 3.8 of the Dutch Aviation Act describes the general prohibition against the execution of flights with an aircraft that (a) is not airworthy or (b) is not provided with a Certificate of Airworthiness.[52] The scope of part (a), the aircraft must be airworthy, is designed to prevent an aircraft that is provided with a Certificate of Airworthiness, however affected by a deficiency or improper modification that actually renders it in a non-airworthy condition, from being used in air operation.

Before an aircraft can be entered in the Dutch national Aircraft Register, the Transport Inspectorate assesses whether that aircraft meets the airworthiness requirements and, whether it is in an airworthy condition.[53] In practice it means that the Inspectorate will assess through inspections and tests on aircraft whether the standards regarding airworthiness are fully complied with. Following a positive aircraft inspection assessment, the Minister of Transport shall issue a Certificate of Airworthiness.

Imported used aircraft could acquire a Certificate of Airworthiness at once, provided that these aircraft hold a valid Foreign Certificate of Airworthiness or an Export Certificate of Airworthiness.

In fact, a Certificate of Airworthiness will be provided to the holder of an aircraft who has done a request. The concept of holder of an aircraft is defined in Article 1.1. sub 1 of the Dutch Aviation Act as: "A person in whose name an aircraft is registered in the national Aircraft Register, as defined in Article 3.3. sub 2, or is registered in a foreign Aircraft Register."

The holder of an aircraft could legally be the owner or the possessor, although this is not essential. An example could be the situation where the holder will lease the respective aircraft for a shorter or longer period. The lessee is in fact the actual holder in that particular case, while the lessor is the one on whose name the aircraft is registered and to be accountable for compliance with airworthiness requirements.

In Dutch aviation legislation it is rather important under whose name an aircraft is registered, but not necessarily who in fact will hold that aircraft. This way someone will always be accountable for the technical condition of a Dutch registered aircraft. Keeping in mind that with respect to the original airworthiness of an aircraft the manufacturer will be responsible, the operator, however, will be primarily responsible for maintaining the expected level of airworthiness, subject to additional national regulations, and will retain full responsibility for the continued airworthiness of its aircraft, even if the operator uses leased aircraft or relies on outside maintenance contractors.

52 Dutch Aviation Act of 18 June 1992 (text as of 21 January 2011).

53 Resolution Aircraft 2008, Article 7: The following Certificates of Airworthiness may be issued: (a) ICAO-standard; (b) EASA-standard; (c) EASA-restricted; (d) Special- or (e) Export Certificate of Airworthiness; (f) Special Permit for operations within the airspace of the EASA member States.

4.17 Maintenance

The owner, respectively the holder of an aircraft for which a Certificate of Airworthiness has been issued, and who intends to carry out commercial air transport operations, is obliged to comply with maintenance, repair and overhaul requirements regarding that aircraft. Maintenance has to be performed by an approved maintenance facility under penalty of revocation of the Certificate of Airworthiness. In case an aircraft is used for commercial air transport, maintenance should be done by a holder of a Maintenance Organization Approval (MOA).

Maintenance consists of repair, overhaul, inspection, modification and rectification of a defect of an aircraft or an aircraft part or a combination of these works, except for any standard pre-flight inspection or equivalent inspections. Overhaul is a maintenance method in which the aircraft or aircraft part will be restored by means of inspection and replacement in accordance with an approved standard with the intent to extend the operational lifetime.

Aircraft maintenance has to be carried out according to a detailed maintenance programme that should be drawn up on the basis of the aircraft manufacturer's recommendations. Deviations from these recommendations are possible in cases such as overhaul or operational limits having been exceeded, without any underlying permission of the aircraft manufacturer or a thorough analysis.

4.18 Transfer of Responsibilities

Leasing of aircraft, dry-lease or wet-lease, is now widely used in commercial air transport. However, lease, charter and interchange arrangements are causing situations in which the owner and holder of an aircraft are not identical. Many aircraft registered in a specific State are leased to air carriers with their headquarters located in different States. This may imply that requirements for operations and maintenance may differ.

When the Chicago Convention was signed in 1944, nobody could have foreseen that international leasing and chartering of aircraft, as we widely know now, would make such significant progress. Government oversight of aircraft, stationed and operated for a long time in a State different from the State of Registry, can lead to complex enforcement problems (this could involve monitoring of flight operations, licensing, maintenance and applicability of relevant national aviation rules) for both the State of Registry and the State of the Operator.

This was more evident when KLM Royal Dutch Airlines was operating in the past, actually in the 1990s, with a number of aircraft registered in the United States, without transfer of aviation oversight competency to the Government of the Netherlands. These special aircraft lease arrangements are nowadays no longer possible. An aircraft can only be entered to the Air Operator Certificate (AOC) if it is registered in the State of the

Operator (national register) or within the Community (Council Regulation (EEC) No 2407/92 of 23 July 1992 on licensing of air carriers).

Arbitrary companies that want to have private air transport capabilities, may be exempted from the requirement to have an AOC.[54] In that particular case, companies do not have to comply with Regulation (EEC) No 2407/92 and consequently could acquire an operating licence for their U.S. registered aircraft. These aircraft only have to comply with the FAR (Federal Aviation Regulations) Part 91, General operating and flight rules, subpart H operations requirements.[55]

In fact there are two distinct situations: commercial air transport with U.S. registered aircraft operated by foreign commercial air carriers is not authorized, while privately operated air transport by companies is allowed with U.S. registered aircraft.

In 1980, during the 23rd Session of ICAO (Montreal, 16 September-7 October 1980), the Assembly agreed on 6 October 1980 to adopt Article 83 bis as an amendment to the Chicago Convention.[56]

> Article 83 bis – Transfer of certain functions and duties –
>
> (a) Notwithstanding the provisions of Articles 12, 30, 31 and 32 (a), when an aircraft is registered in a contracting State is operated pursuant to an agreement for the lease, charter or interchange of the aircraft or any similar arrangement by an operator who has his principal place of business or, if he has no such place of business, his permanent residence in another contracting State, the State of Registry may, by agreement with such other State, transfer to it all or part of its functions and duties as State of Registry in respect of that aircraft under Articles 12, 30, 31 and 32 (a). The State of Registry shall be relieved of responsibility in respect of the functions and duties transferred.
>
> (b) The transfer shall not have effect in respect of other contracting States before either the agreement between States in which it is embodied has been registered with the Council and made public pursuant to Article 83 or the existence and scope of the agreement have been directly communicated to the authorities of the other contacting State or States concerned by a State party to the agreement.
>
> (c) The provisions of paragraphs (a) and (b) above shall also be applicable to cases covered by Article 77.

---

54 Dutch Aviation Act, Chapter 4.

55 FAR Part 91 subpart H: Foreign Aircraft Operations and Operations of U.S.-Registered Civil Aircraft Outside of the United States; and Rules Governing Persons on Board (POB) Such Aircraft.

56 ICAO A 23-WP 16 EX/4: Amendment to the Chicago Convention with respect to problems arising out of lease, charter and interchange of aircraft in international operations (proposed Article 83 bis), 21 April 1980. The amendment was ratified by the Netherlands on 5 November 1981. Article 83 bis entered into force on 27 June 1997. (*See*: <www.icao.int/icao/en/leb/83bis.pdf>.)

Article 83 bis grants governments in bilateral agreements to delegate oversight on leased or chartered aircraft to another contracting State.

> Although the Chicago Convention allocates to the State of Registry certain functions which that State is entitled to discharge, or obligated to discharge, as the case may be, the Assembly recognized, in Resolution A 23-13, that the State of Registry may be unable to fulfil its responsibilities adequately in instances where aircraft are leased, chartered or interchanged – in particular without crew – by an operator of another contracting State and that the Convention may not adequately specify the rights and obligations of the State of an Operator in such instances until such time as Article 83 bis of the Convention enters into force. Accordingly, the Council urged that if, in the above-mentioned instances, the State of Registry finds itself unable to discharge adequately the functions allocated to it by the Convention, it delegates to the State of the Operator, subject to acceptance by the latter State, those functions of the State of Registry that can more adequately be discharged by the State of the Operator. It was understood that pending entry into force of Article 83 bis of the Convention, the foregoing action would only be a matter of practical convenience and would not affect either the provisions of the Chicago Convention prescribing the duties of the State of Registry or any third State. However, as Article 83 bis entered into force on 20 June 1997, such transfer agreements will have effect in respect of those contracting States which have ratified the related protocol (Doc 9318) upon fulfilment of the conditions established in Article 83 bis.[57]

Moreover, States were already able to delegate oversight before Article 83 bis entered into force, but from that date this specific delegation has been widely recognized by other contracting States. The delegated functions are: monitoring compliance with rules of the air, aircraft on-board radio-communication equipment, Certificates of Airworthiness and certificates of competency and licences of crew members.[58] The transfer must be registered with the ICAO Council and made public or it should be directly communicated to the relevant authorities of the other contracting States concerned as party to the respective agreement. Otherwise the transfer will have no legal force.

---

57 ICAO Annex 8 Airworthiness of Aircraft, Part II Procedures for Certification and Continuous Airworthiness (Note).

58 Article 12 (Rules of the Air), Article 30 (Aircraft Radio Equipment), Article 31 (Certificates of Airworthiness), and Article 32a (Licenses of personnel) of the Chicago Convention.

4.19 Bottom-Up and Top-Down Approaches

In this chapter some weaknesses in the Chicago Convention system are exposed.

In the first place it is clear that although ICAO may establish safety standards regarding (continuous) airworthiness and maintenance of aircraft, it is unable to act in a coercive manner. Instead, ICAO looks towards a transition to a more results-oriented and performance-based organization, in line with fundamental changes in global aviation and the evolving requirements of all constituents.[59]

Sovereign States simply do have full control on airworthiness of aircraft and flight operations. If all States adhere to this rule everything would be in good order in this respect, but the phenomenon "risky carriers" has shown that it cannot always be trusted. Only relatively recently has this fact become clear. The reason can be found in the fact that apparently within ICAO, people have been too preoccupied with the fiction that States naturally comply with the provisions of agreements.[60]

In 2000 the Civil Aviation Authorities of the Netherlands and the Netherlands Antilles have been subject to an audit. The results of this audit are specified in an ICAO USOAP audit report. Some important findings regarding the flight operation inspection programme as hitherto carried out by the Dutch authorities were:

The current inspection policy at that time (2000) did not provide for oversight on all categories of inspection and did not meet the minimum frequency of inspections, as recommended by ICAO guidelines. The system did not adequately reflect the magnitude, scope and complexity of air carriers' operations. The flight inspection group has only performed 18% of the planned inspections for the year 2000, although there already existed a limited inspection programme. In response to these findings, the Dutch Government has taken the necessary measures. In 2003, in an ICAO USOAP follow-up audit, these actions of the Dutch Government were characterized as sufficient. The next USOAP audit was planned for 2007 (actually performed in 2008).[61]

On the basis of international adopted agreements, the State of Registry is responsible for oversight regarding the airworthiness of aircraft, except for the possibility of delegation if the State of Registry finds itself unable to adequately discharge this function, allocated to it by the Chicago Convention. The system of the Chicago Convention assumes that contracting States will rely on the fact that these States shall actually and thoroughly carry out such oversight. Unfortunately this is not always true. Some incidents involving

59 ICAO USOAP Report 2007, Comprehensive Systems Approach, at 3.

60 Anthology Consultation Round regarding Safety Policy on Civil Aviation (summary), July 1995, NVB 047, at 12. Also Civil Aviation Department-Public Relations Officer H. Scholten in Journal *HP/De Tijd* of 16 February 1996.

61 *See*: *Luchtvaartveiligheid in Koninkrijk doorstaat ICAO-audit* (Aviation Safety within the Kingdom of the Netherlands succeeds ICAO-audit), luchtvaartnieuws.nl, 19 May 2008. *See also*: ICAO USOAP Audits within the EU & EASA MS 2008, Overview of issues addressed to EASA, 28 November 2008.

foreign aircraft at Schiphol international airport, in the early 1990s, gave reason to doubt the airworthiness status of aircraft belonging to some foreign air carriers.

Initially, the Minister of Transport considered the ICAO system to be performing adequately and that one could simply rely upon safety oversight conducted by other contracting States. However, from audits performed by the U.S. FAA in the 1990s, it became clear that safety oversight was well below standards in a number of contracting States, contributing to hazardous air carrier operations. In the Netherlands, but in other European States as well, the aviation authorities became aware of a new phenomenon: "risky carriers".

When the FAA started an investigation concerning the implementation of ICAO standards in different States, the overall conclusion was that implementation was not done at all or has been incomplete.

The United States and the European Union themselves have supplemented the imperfections in the ICAO system. The JAA (or ECAC) member States, including the Netherlands, have chosen for the institution of so-called Safety Assessment of Foreign Aircraft (SAFA) teams.

SAFA teams are inspection teams performing ramp inspections on foreign civil aircraft at civil airports and civilian sections of military aerodromes. Aircraft are subject to a thorough, although time-limited, inspection by a SAFA team. If defects are detected, the team has the authority to ground an aircraft and to take necessary measures to ensure that the air operator will bring the aircraft in an airworthy condition in order to acquire permission for departure (bottom-up approach). In contrast to the European Union, the United States will first assess to what extent the national aviation legislation of a visiting foreign air carrier is able to provide the opportunity to meet the minimum requirements of the ICAO, then assess the way the relevant rules are enforced, and more in particular how air traffic permits to air carriers and routine inspections are organized (top-down approach).

The American approach considers and assesses whether a foreign government is able to give substance to aviation safety oversight. An air carrier from a State with a poorly organized safety oversight system is not allowed to fly to the United States. With respect to the EU SAFA programme, its aberrant approach is especially focused on the air carrier as being the central point, to a greater extent than the oversight authorities of the State of the Operator.

The ICAO has no authority to intervene regarding the airworthiness oversight of contracting States. Audits conducted under the USOAP and compulsory for contracting States should uncover any defects. However, not every contracting State will have sufficient and adequate knowledge and resources to remedy deficiencies in the field of aviation safety oversight. On request of contracting States, ICAO has the ability to support those States in setting up a full and effective safety oversight system. ICAO itself collects data from various (USOAP) safety oversight audits that enables it to:

a. determine the level of aviation activity: by State, by region, by a select group of States, or globally;
b. determine contracting States' level of implementation of SARPs for each of the safety-related ICAO Annexes, based on compliance checklists completed and submitted by States and audits results;
c. determine the percentage of lack of effective implementation broken down by the eight critical elements (CEs) of a safety oversight system: by State, by Region, by a select group of States, or globally; and
d. determine the types of difficulties experienced by contracting States in establishing an effective safety oversight system in each of the eight areas audited:
   i. primary aviation legislation and civil aviation regulations;
   ii. civil aviation organization;
   iii. personnel licensing and training;
   iv. aircraft operations;
   v. airworthiness of aircraft;
   vi. aircraft accident and incident investigation;
   vii. air navigation services; and
   viii. aerodromes.[62]

In Europe the first measures to enhance safety in civil aviation were reflected by the adoption in 1996 by the European Commission of a so-called Communication entitled: "Defining a Community Aviation Safety Improvement Strategy", proposing to enhance safety in civil aviation by ensuring that aircraft comply fully with the safety standards defined by the Chicago Convention, to be established by a system based on a bottom-up approach. This type of approach was already common practice in the voluntary ECAC SAFA programme.

> Safety Assessment of Foreign Aircraft.
>
> As a consequence, in its action plan, the Commission announced its intention "to present … a proposal for a Council Directive formalizing the SAFA procedure for the assessment of foreign air carriers and the related cooperation mechanism to share and analyze information and draw conclusions. This directive should also include an obligation on member States to ground aircraft found or suspected to be dangerous".[63]

In 2001, the Commission proposed to enhance air safety globally and to adopt a Directive (2004/36/EC) on the safety of third-country aircraft using Community airports.[64]

62 ICAO USOAP Report 2007 Comprehensive Systems Approach, at 4.
63 SEC (96) 1083 final, Defining a Community Aviation Safety Improvement Strategy, 12 June 1996.
64 *See* COM (2001) 390 final.

Effective enforcement of international safety standards in the Community will be established by harmonizing the rules and procedures for ramp inspections of third-country aircraft, excluding both State aircraft as defined in the Chicago Convention and aircraft not exceeding a maximum take-off mass of 5,700 kg and not engaged in commercial air transport. Rules have been established regarding ramp inspections, collection, exchange, protection and dissemination of information and grounding of aircraft.

In this context, the SAFA programme could be considered a sophisticated and well-functioning filtering system, but at the same time a rigorous method, if applied on a wide scale, to reveal deficiencies not only regarding aircraft, but also air carriers and eventually relevant governmental safety oversight systems.

The programme works as a big stick or discouragement to air carriers in order to avoid incurring an operating ban within the European Union. The imposition of an operating ban is intended to ensure the safety of people, either in the air or on the ground, not only within the European Union but on all routes still operated by the potentially ill-fated air carrier. Moreover, inclusion of an air carrier means strict tracking of that carrier and its oversight authorities with respect to corrective actions regarding identified deficiencies and compliance with minimum safety standards as well as profound supervision. To enhance aviation safety in any third country that displays the demand and thoroughness to move forward in this field, the SAFA programme incorporates the ability to assist when necessary.

4.20 Concluding Remarks

In the 1990s, globalization and deregulation of civil aviation, the fastest growing form of transport, induced intensified air traffic to airports located in the European Union. Third-country aircraft, to a large extent improperly operating in full compliance with international safety standards, were increasingly part of the total air traffic within Community airspace and had an unacceptable impact on the European Union. Although the accident rate in the last decade was low in comparison with many other human activities, the number of aviation accidents had increased, which fuelled a public perception of a decreasing level of safety.

To ensure the safety of European citizens travelling by air worldwide or living around major airports, the European Commission elaborated a Community Aviation Safety Improvement Strategy that was adopted and transmitted to the Council on 12 June 1996. To further verify that third countries and their air carriers shall actually apply internationally agreed aviation safety standards and to assist them in doing so when necessary, the European Commission proposed in 2001 to the European Parliament and the Council to adopt a Directive on the safety of third-country aircraft using Community airports, so that appropriate information could be gathered and actions taken by the Community and its member States to rectify dangerous situations caused by third-country aircraft.

The SAFA programme, established by ECAC, adopted and further developed by the European Commission, provides European States with a surveillance tool so that they are made aware of identifiable deficiencies and can act in consequence, mainly by random ramp inspections of foreign aircraft. Findings are categorized to classify the seriousness of the shortcomings in incremental order.

SAFA inspections have no direct connection with the assessment of a State's oversight capability. However, as a bottom-up approach it is considered that, whilst certainly not replacing safety oversight system assessments or audits, it contributes upstream of such assessments by drawing attention to possible deficiencies in a State's oversight system.

The Chicago Convention and its Annexes allocate the responsibility for aviation safety to individual States. Each contracting State bears responsibility for the safe design and continuing airworthiness of aircraft, for safe operations and for safe air traffic flow within its own national airspace, including the necessary ATM and airport infrastructure. If these obligations are not fully respected by contracting States, air safety deficiencies may arise. In this context it is important that ICAO standards are implemented and enforced throughout the world.

The air carrier has a responsibility for the safe conduct of its operations and also to comply with any regulations, internationally agreed and implemented, or national regulations and rules that the State may promulgate. However, too stringent rules may have a paralyzing effect on all stakeholders.

Poor safety oversight and underperforming air carriers, causing unsafe conditions, can be penalized by inclusion of these air carriers in the EU blacklist. Poor State safety oversight can be pushed in the right direction by EU member State aviation authorities, EASA and with the assistance of various non-governmental organizations. Air carriers must, in fact, immediately demonstrate compliance with international safety standards to be lifted from the ban. A method to circumvent this problematic issue could be leasing of aircraft, which has come into wide usage. Unless suitable arrangements are made between the States involved, a lease may create complex legal, safety, enforcement and practical problems for both the State of Registry of the aircraft concerned and/or the State of the Operator because of possible uncertainty concerning which party is responsible for the safe operation and airworthiness of the aircraft, and which State's regulations are applicable. Air carriers operating with leased aircraft because of an operating ban on their own fleet, could face a bewilderment of operations, registration, maintenance, licensing and regulations, if not already in the process of evading responsibilities.

5 Eliminating the Phenomenon of Risky Carriers

5.1 General

There are air carriers in the world known for non-compliance with minimum safety standards as agreed within the ICAO structural consistency, for example because of flight operations conducted under known unsafe conditions, reduction of aircraft maintenance expenditure or insufficient training of flight crew members (*e.g.* lack of sophisticated full flight simulators). These air carriers are also referred to as "risky carriers".[1]

A risky carrier is defined as: "An air carrier whose aircraft, crew members and/or flight operations, based on a well-founded judgement, are not complying with the applicable ICAO safety standards".

However, if an air carrier in a specific State does not comply with the minimum safety standards, this poor performance will not be just the fault of that carrier, but will most probably be caused by the fact that the government of that State lacks laws, regulations and resources necessary to support proper oversight of that air carrier in accordance with applicable minimum international standards.[2] Indeed, the aviation authority of the State of Registry is fully responsible for the safety oversight of its own air carriers, as can be abstracted from Article 31 of the Chicago Convention. It may be assumed to be known that a number of States, unfortunately, are less able to establish effective and independent aviation safety oversight.

Regarding the European region, the European civil aviation framework has changed dramatically over the previous decades. Most probably the biggest change was the demise of the former Union of Soviet Socialist Republics (USSR) into many Republics and independent States, both in part belonging to the European region as to the Asian region.

In addition, deregulation and open skies policies, and particularly the expansion of the European Union with various Eastern European countries, generally lagging behind the modern western civil aviation culture, caused adjustment challenges.

---

1 Note concerning problems arising from risky carriers in civil aviation, Letter of the Dutch Minister of Transport to the Chairman of the Second Chamber of Parliament of 10 November 1995, Parliamentary Documents II 24 400 XII, No 25.

2 L.H.D. Bakker, 'Risky Carriers or Risky States?', *in* H.M. Soekkha (Ed.), *Aviation Safety: Human Factors – System Engineering – Flight Operations – Economics – Strategies – Management*, VSP BV, Zeist 1997, at 275.

The break-up of the Soviet Union created major aviation management problems that the rest of the world, particularly Europe, had to cope with. Rather than one centralized aviation authority, as many as 18 different authorities were formed. ICAO's Paris Regional Office accredited list increased during this transition from about 25 States (of which 12 were EU member States at that time) to 51 States.

It obviously meant that all these new ICAO member States, States that had little, if any, aviation management experience, were to sign the instruments of acceptance to the Chicago Convention and as a logical consequence shortly after, to implement its provisions.

Many decades ago, after the Second World War, a similar situation developed in other regions of the world when many former colonies became independent States. The aviation experiences of many of these former colonies are closely associated with those of their European colonial rulers. In fact, the present aviation organizational structures and aviation rules and legislations of these former colonies are basically copies or adaptations of the established European structures and adopted legislations, which were applicable at the time when these colonies acquired their independence.

Most of these (developing) States, apart from some rapidly emerging States, still have, after years of independence, rather weak, poorly functioning civil aviation management structures, inadequate aviation infrastructures and air carriers. This situation can be explained by the fact that well-established States in the field of aviation have historical norms for their aviation safety cultures and have adequate built-in layers of professional protection and, in addition, sufficient resources, contrary to the often typical inherited bureaucratic administration system of developing States.

This developing States dilemma was recognized by the western world and required, at the governmental regulatory oversight level, substantial technical and administrative aid through bilateral, even multilateral, agreements. A number of these independent States are still in aviation safety disorder, despite the efforts put in by ICAO and other organizations and institutions. Even emerging States, mostly former colonies, economically boosted nowadays, are facing the struggling race between, on the one hand, their ancient structured civil aviation authorities and, on the other hand, the rapidly extending massive air transport market and production levels involved and consequently the greater complexity of aviation safety issues. Most, if not all, resources to address this imbalance are available in Europe.

5.2 Safety Oversight and Risky Carriers

When the FAA, the United States civil aviation authority, did research in the 1990s in the field of compliance with ICAO regulations by foreign air carriers operating to the United States, the result was that only one-third of the ICAO member States fully respected and complied with all ICAO regulations. Since then, the FAA has imposed assessments on

States already having traffic rights or States applying for these rights to operate to the United States.

These screenings are conducted through the International Aviation Safety Assessments (IASA) Programme, thereby assessing the extent to which national legislation is able to provide the opportunity to meet the minimum requirements of ICAO, to set out the procedure for how the rules are enforced, in particular the authorization of air carrier operating certificates and licences and in what manner routine inspections are organized.

The assessments determine whether or not the State has a civil aviation authority in place and the extent to which that authority ensures that operational and safety procedures are maintained by its national air carriers.

According to ICAO Annex 6, it is mandatory that contracting States have a system to monitor the flight operations of their air operators/air carriers in international air transportation. The purpose of Annex 6 is to contribute to the safety of international air navigation by providing for safe operating practices and to contribute to the efficiency and regularity of international air navigation by encouraging ICAO's contracting States to facilitate the passage over their territories of commercial aircraft belonging to other countries that operate in conformity with these criteria.

Such a monitoring system shall normally consist of a fully functioning civil aviation authority to which the State will delegate, *inter alia*, its aviation safety monitoring activities and which is capable of ensuring compliance with operational and safety procedures by its air carriers.

A State shall comply with the ICAO standards (category 1) or rather not comply (category 2). Category 2 States are not sufficiently able to monitor air carriers due to inadequate legislation or expertise. If the civil aviation authority of a State with no existing air carrier service to the United States does not have an acceptable level of safety oversight, its application for economic authority will be denied, and subsequently commercial aircraft from that State, in principle, will not be admitted. The FAA will undertake a reassessment of the civil aviation authority after evidence of compliance with ICAO requirements has been received. However, in case category 2 State commercial aircraft are wet-leased to an authorized U.S. air carrier or to an air carrier from a category 1 State, then these aircraft are admitted to the U.S.[3]

If after an assessment the civil aviation authority of a State with an existing air carrier service to the United States does not meet the minimum standards, the FAA gives that civil aviation authority a category 2 rating, which means that air carriers from the assessed State

---

3 A wet-lease is a (normally short-term) leasing arrangement whereby one air carrier (lessor) provides an aircraft, complete crew, maintenance and insurance (ACMI) to another air carrier (lessee) which pays by hours operated. The lessee provides for fuel, pays taxes, airport fees and any other duties. There are various leasing constructions depending on different conditions: Dry-lease (operating or finance), Wet-lease (in-out) and Damp-lease.

cannot initiate new services and are restricted to current levels of any existing service to the United States, while corrective actions are underway. The FAA formally requests consultations with the relevant civil aviation authority, which in fact might be a delicate issue.

For many years the international aviation community considered ICAO Annex 6 as too inviolable for probing from outside States because of sovereignty aspects if full adherence was sought from certain consult States. The principle of respect with regard to national sovereignty is extremely important in international law. Rules of international air transport, as laid down in the ICAO Annexes, are no different.[4]

Risky air carriers, considered as carriers failing to comply with international safety standards, are very often the result of inadequate governmental oversight, *in casu*, States that are still unable to meet their aviation safety management obligations as ICAO member States. These States often lack the fulfillment of safety oversight obligations, identified as four basic areas: certification, surveillance, enforcement and licensing, owing to the absence of critical safety oversight elements as updated and enforceable primary aviation legislation and regulations, expertise, training, guidance procedures and monitoring the status of home-basedair carriers and registered civil aircraft. The essence is to create an efficient aviation safety management through an adequate institutional set-up, and adequate aviation legislation in a State.

There is no reason to doubt the fact that a State's safety oversight system must comprise appropriate developed and established procedures and organizational processes that generally enable a contracting State to satisfactorily implement international SARPs and accepted safe practices. Where proper safety oversight systems are not established, the potential for failures in the system is significant.

A particular aviation-related deficiency may or may not be a direct causal factor, but its contribution to a failed defence system for the safe operation of an aircraft is not in doubt. Especially in that environment, a poorly monitored air carrier could drift into the risky carrier status. Safety oversight audits, assistance and safety assessments enable the identification of shortcomings and possible failures well before they emerge as a threat to aviation safety, and thus help to strengthen the defences within the overall aviation system, in particular the safety oversight systems of underperforming contracting States.

A well-organized governmental organization and infrastructure in the field of civil aviation will normally have a positive reaction on air carriers. Through adequate inspections and safety enhancement assessments these carriers are forced to strive for a balanced organizational structure, which will include a key position for safety management systems, a prerequisite to improve air carrier safety standards.

---

4 Organization for Economic Co-operation and Development (OECD), *Globalization, Transport and the Environment*, 2010 (ISBN 978-92-64-07291-6), at 251.

Pressing requirements of safety assessments from the FAA (Categorization) and the European Union (Risky Carriers and EU blacklist) made it obvious that their evaluations were only part of an overall safety need of civil aviation worldwide. In fact, all ICAO Annexes, not only those often called "safety Annexes 1, 6 and 8", are part of an overall safety environment as they were designed to be an integrated, internationally accepted, aviation safety umbrella.

The aviation administration of any ICAO member State should therefore be capable of implementing, managing and overseeing all 18 ICAO Annexes and associated procedures as well as guidance materials. However, the approaches used so far for this purpose often fell short of achieving their goals. One has only to look at the results of the FAA and ICAO safety assessment efforts in the distant past.

These early assessments covered mainly the ICAO Annexes 1, 6 and 8, while the European Union specifically in the beginning was focusing on Annex 8 and to a lesser degree on Annexes 1 and 6. In some cases, compliance with other ICAO Annexes was taken into consideration with the aim of promoting safety by improving the implementation of safety-related uniform SARPs.

Comprehensive aviation safety (and security) assessment programmes so far have drastically contributed to aviation safety in a broader context.[5] Aviation safety assessments and security audits, in combination with aviation safety management assistance to third country civil aviation authorities, will eventually strengthen governmental aviation oversight, implementation of internationally adopted safety standards, preferably all ICAO SARPs, and thus the capability of reducing and eventually eliminating the continued existence of risky air carriers in these countries.

5.3 SAFETY OVERSIGHT AND SAFETY MANAGEMENT SYSTEMS

Safety oversight of national air carriers must be provided by a State's authority, which must be well equipped, organized and experienced. Failure to provide effective and adequate aviation safety oversight may be due to political motives, economic protection, inefficient organization structures or deregulation. This particular situation may lead to indifferent processing of deficiencies in operation and maintenance of these national air carriers. Factors that worsen this deplorable state of affairs are widespread leasing and sub-leasing of aircraft to substandard carriers and self-government of former dependent States.[6]

5 Regarding State safety oversight: IASA, SAFA and USOAP. Regarding flight operational safety: FAA's Aviation Safety Information Analysis and Sharing (ASIAS) programme, Flight Operational Quality Assurance (FOQA) programme and Aviation Safety Action Program (ASAP). Regarding global aviation security: ICAO Universal Security Audit Programme (USAP).

6 The demise of the former USSR created major aviation management problems. Instead of one centralized aviation authority, as many as 18 different authorities have been formed.

Effective aviation safety oversight requires a number of implementing tools, which include the following: availability of primary aviation legislation and the provision of effective aviation law consistent with the environment and complexity of the State's aviation industry; specific operating regulations and the provision of adequate regulations to address, at a minimum, ICAO Convention requirements and Annex standards; a civil aviation structure and the provision of a civil aviation organization, led by an appropriate empowered executive, with the necessary technical supervisory and support personnel, including the provision of adequate resources; qualified technical personnel to inspect the aviation community's activities to the level of complexity required; documentation and guidance material, the provision of sufficient technical guidance to allow the technical and support personnel to perform their functions in a standardized manner; licensing and certification activities, the provision of processes used to ensure that the aviation community continues to function at minimum required levels of competency and safety; and a system for the resolution of safety concerns, the provision of processes and methodology to facilitate the resolution of safety issues, which includes the ability to take action to penalize, suspend or revoke an individual's or organization's authority to participate in aviation activities. It is the effective implementation of these critical elements that constitutes a safeguard against latent conditions and failures.

A very important safety aspect considered to be one of the major elements of a State's aviation safety oversight system, is the establishment of a State safety management system, which plays a major role as a safety implementation tool available to a State's civil aviation administration. A safety management system (SMS), at the government level but at the same time at the industry level, is a systematic, explicit and comprehensive process for managing safety risks.

A safety management system could be defined as a dynamic risk management system based on quality management system (QMS) principles in a structure scaled appropriately to the operational risk, applied in a safe culture environment.[7]

> One of the purposes of a Safety Management System is to improve the safety performance, and therefore reduce the exposure to risk of having an accident. It is not focused on the safety record per se. Quality systems are focused on continuous improvement also, but through improving the production record rate. This is another source of confusion between the two management system

---

7 A.J. Stolzer, C.D. Halford & J.J. Goglia, *Safety Management Systems in Aviation*, Ashgate Publishing Company, Burlington, VT 2008, at 19.
A Safety Management System is a systematic and explicit approach defining the activities by which safety management is undertaken by an organization in order to achieve acceptable or tolerable safety (definition by EUROCONTROL Safety Regulatory Requirements – ESARR3). The FAA definition: SMS is the formal, top-down business approach to managing safety risks, which will include a systematic approach to managing safety, including the necessary organizational structures, accountabilities, policies and procedures.

> concepts; improving a safety record, is not the same as improving safety performance. There are many aviation companies that have extremely good safety records, but are operating with risky behaviour or inadequate organizational structures, and have just not had an accident yet. A good safety record, just like a good quality record, does not guarantee safety.[8]

> When effectively combined, the SMS and QMS assure us of safe and quality products. Stated simply: SMS needs a quality management system because SMS requires documented, repeatable processes. The quality management system provides the strong foundation upon which an SMS is built. Aviation safety's QMS will be the springboard to our SMS.[9]

A State's safety oversight function is a specific part of a State's Safety Programme. The objectives of a State's safety oversight function are satisfied through administrative controls like inspections, audits and surveys carried out on a regular basis by civil aviation authorities to identify hazards and mitigate risks. However, the increasing complexity of both aircraft and air carriers engaged in civil aviation has been accompanied by corresponding changes in the number of requirements and the demands put upon the regulating and oversight authorities. The result is that civil aviation authorities can no longer obtain an adequate picture of a complex aviation company operation without considerable and progressive enlargement of their resources.

The scale of increases that would be required are not affordable by the industry as a whole, nor would such an increase be the best response to the problem. This situation applies not only to ICAO standards-compliant States, but in particular to States with risky carriers. Starting from the fact that government authorities continue to be responsible for overall surveillance and for drawing up the regulations that form the basis on which a safe operation is built, and the fact that air operators remain responsible for the safety of their operations and compliance with the requirements, the concept of quality assurance is a substantial contribution to aviation safety. The emphasis in regulation is therefore likely to change regular inspections by the aviation authority to a system whereby operators self-audit their own operation against a clearly defined quality system. The availability of an effective internal auditing system, in other words an effective aviation company quality assurance system, will ensure that complacency has not settled in as a result of conducting routine activities.

The authority's role eventually shifts to the verification of operators' compliance with their quality and safety management systems and ultimately with the safety requirements. The development of internal audit and evaluation programmes relies heavily upon

8 FAA, Safety Management System SMS Explained, 30 October 2009.
9 Statement of N.A. Sabatini, FAA Associate Administrator for Aviation Safety, 2005.

AOC holders to continuously monitor and audit their operations to ensure that they are safe and conform to the requirements and the operator's standards. The quality system must include a Quality Assurance Program (QAP) that contains procedures designed to verify that all operations are being conducted in accordance with all applicable requirements, standards and procedures.

Identifying air carriers that have a lower safety record than is accepted by ICAO, and/or which are not fully complying with internationally agreed safety standards, is the primary objective of safety assessment programmes such as SAFA.

But identifying risky carriers in general terms means almost inevitably uncovering more complicated problems, *in casu*, defects regarding State safety oversight, the absence of adequate safety and quality management systems as well as poor implementation of standards and requirements. An air carrier is obviously not the aviation safety authority of the State. Moreover, air carriers do not sign the adherence to the Chicago Convention. As a commercial enterprise their vested interests are rather different from those of government authorities.

Although aviation safety should be a prerequisite for subsistence for an air carrier, its safety interpretation could be very subjective and could significantly differ from that of the international auditing organization or even its own State oversight authority assessing that particular carrier. However, practice has repeatedly and indisputably shown that a State, where a risky carrier originates, does not exactly respect applicable aviation safety requirements.

Transfer of expertise, support of monitoring safety oversight performance and safety regulatory auditing as well as providing training and expanding adequate resources and communication will eventually contribute to proper working safety and quality management systems and oversight inspectorates, resulting in a drastic reduction in the number of underperforming States and, consequently, their national risky carriers.

A rather recent phenomenon in an attempt to enhance air transport operational safety can be attributed to code-sharing. The increasingly popular code-share alliances will, from both economic and safety considerations (reputations), ensure that certain participating air carriers will meet an accepted level of safety.[10]

5.4 REFLECTION

An air carrier organizational structure with its complex and potentially hazardous but professionally responsible flight operation needs a comprehensive and adequate safety management system to monitor previously unexpected and unforeseen risks, to identify

10 U.S. Department of Transport Code-share Safety Program (2000). This programme requires U.S. air carriers to conduct safety audits on their foreign code-share partners as a condition of code-share authorization.

hazards and to control and mitigate acceptable risks. A safety management system is a systematic, explicit and comprehensive approach to managing safety risks, including analyzing and addressing identified shortcomings to be able to improve the safety level of the organization.

It represents a system for the management of safety within the industry, including the organizational structure, responsibilities, procedures, processes and provisions for the implementation of the relevant operations by an operator, approved maintenance organization (AOM), aviation training institute, etc., which provide for control of safety and order at all levels of activities, as applicable.

No human activity, like commercial aviation, can be guaranteed to be absolutely safe. In other words, there will always be inherent risks. An air carrier's safety management system, acceptable to the State of the Operator, takes a proactive approach to anticipate and address safety issues before they materialize and possibly lead to aviation incidents and accidents.

> Safety is the state in which the risk of harm to persons or of property damage is reduced to, and maintained at or below, an acceptable level through a continuing process of hazard identification and risk management.[11]

Every ICAO member State is required to provide for the safe and efficient operations of aircraft within its airspace. Some of the general safety obligations include ensuring provisions for aircraft operations, airworthiness and maintenance, licensing, air traffic management and supervision of national air carrier operations. It is clear that the State of the Operator has major obligations to ensure air transport safety, in general, and, more specifically, the safe transport of passengers and cargo on aircraft operated by the State's air carriers.

A common problem faced by a number of ICAO member States relates to the establishment of an adequate civil aviation organization appropriately empowered to regulate, control and supervise aviation activities – an organization with the right number of qualified and experienced personnel, adequate funds to enable it to manage its responsibilities and duties in the most effective and efficient manner, along with the necessary tools and equipment, including documentation, guidance material and, most importantly, the authority to conduct the business of a civil aviation authority.

In a number of ICAO member States, several of the requirements that would signify an appropriately established civil aviation organization are missing and, consequently, there is no question of the presence of a truly effective system for regulating and supervising civil aviation activities. In several cases, lack of adequately qualified and experienced

11 Safety Management System, Bangalore International Airport Limited, September 2007, at 8.

technical staff still constitutes a major problem experienced by States, including lack of training policies and programmes for inspectors, written procedures, documentation and equipment, as well as funding.

Civil aviation administrations that fail to have well-structured, active and effective organizations, capable of adequate and fully compliant safety oversight, will have a seemingly insurmountable negative effect on their national air carriers. The result will be underperforming air carriers, known as risky carriers.

This phenomenon emerged in Europe and Asia due to the demise of the former USSR, creating different major aviation management problems both at the authority level and with regard to a large number of poorly equipped air carriers of independent States, all sharing civil air traffic in an unfamiliar competitive environment, adversely affecting aviation safety. De-colonization in different parts of the world established independent States and air carriers, though most of them were still in aviation safety despair. These "ICAO rules deficient States" will be supported by organizations like EASA, based, among other things, on relevant SAFA assessments, to properly execute their ICAO responsibilities, and, as a consequence, to conduct proper regulatory oversight on their ailing air carriers.

The proposition "Strong States would also make available experts and teams for advice and assistance to the less fortunate States" is a key issue of EU bilateral and multilateral agreements, working arrangements and technical cooperation with less-developed countries and third country regulators to ensure the highest level of civil aviation safety. EASA, as the European executive coordinating civil aviation authority, states in an explanatory note that international cooperation is closely linked to rulemaking as it covers both the development of international standards and the conclusion of international agreements or arrangements that become binding on the European Union or EASA. The work of EASA, in terms of international cooperation, comprises assistance to the European Union and its member States in harmonizing rules and promoting European standards at the international level on a multilateral basis.

Individual EU member States and European organizations such as EUROCONTROL are working in close collaboration with EASA on systematic cooperative efforts with third countries. This collaboration rests on the Common Market requirement that the Community speaks with one voice at the international level.

Powerful European national carriers are cooperating with less-developed-country air carriers to assist in standardizing and rationalizing management, maintenance and flight operations, mostly by joint ventures or alliances. However, these poorly performing air carriers must be very reluctant in view of possible outflanking by powerful major air carriers.

Regarding bilateral cooperation, EASA also works at facilitating the free movement of European products and services worldwide. This can be achieved by assisting third-country regulators certifying European products and service providers. Reciprocally,

European certificates can be issued on the basis of third-country certificates when there is sufficient confidence in the regulatory system of partner authorities to use their findings.

Bilateral agreements or working arrangements have to be concluded by the European Union or EASA, depending on the nature of the subject to be covered. EASA therefore assists the European Commission in negotiating agreements that have to bind the European Union and its member States, or EASA negotiates directly regarding the arrangements covering its own field of activities.

Regarding technical cooperation, EASA shall assist less-developed countries in improving their regulatory capabilities to ensure the highest level of aviation safety, because European citizens are travelling all over the world and third country aircraft fly to EU airports. In this domain EASA also works with the Commission in the development and implementation of significant cooperation projects in Asia and Central America.

6 Aviation Safety Oversight Programme

6.1 Introduction

Because of the inadequate implementation of safety standards in some contracting States, as was determined by the FAA, ICAO decided to establish the Safety Oversight Program (SOP).[1] This programme was adopted by the contracting States on the ICAO General Assembly Session in October 1995.[2]

This audit programme aimed at ensuring the effective implementation by the contracting States of the Standards and Recommended Practices that were included in the ICAO Annexes (Annex 1 Personnel Licensing, Annex 6 Operation of Aircraft and Annex 8 Airworthiness of Aircraft).

The main function of the ICAO programme was to conduct voluntary assessments of safety oversight performed by contracting States, with the aim of providing further guidance and necessary technical assistance, if applicable, to the States concerned to enable the implementation of SARPs and associated ICAO procedures. Such a specific evaluation is an audit that assesses the extent to which a contracting State actually corresponds to internationally agreed standards.

During this audit an assessment will take place of whether the regulatory authorities do have access to legal tools, resources, personnel and expertise to meet their international responsibility as supervisor regarding aviation safety oversight.

In order to help contracting States to take the necessary remedial measures to correct deficiencies found during a safety audit, ICAO will prepare, in consultation with and with the consent of the State concerned, an approved action plan. This plan provides the basis for the restoration of full compliance with the ICAO Annexes.[3]

1 Safety oversight is a process by which States ensure implementation of SARPs at the domestic level through the enactment of primary aviation legislation and the establishment of an oversight agency, typically in practice a Civil Aviation Authority (CAA).

2 ICAO Assembly Resolution A29-13 On Improvement of Safety Oversight recognized that many contracting States may not have the regulatory framework to carry out the implementation of the minimum requirements of the Convention on International Civil Aviation and it Annexes. It also recognized that the effective implementation of the safety Standards required effective oversight by the contracting States. The ICAO Safety Oversight Program (SOP), as approved by the council, was endorsed by the 31st Session of the Assembly in 1995.

3 ICAO Assembly Resolution A32-11 Establishment of an ICAO Universal Safety Oversight Audit Programme, ICAO Doc 9730 (1998).

The original Safety Oversight Program showed a number of shortcomings. It was hampered in its effectiveness because of its dependence on voluntary contributions from contracting States, as well as by continuing funding problems within ICAO. Since the programme was based on voluntary participation, it could not always be applied to situations where there was greatest need. In November 1998, during the 32nd General Session of the ICAO Assembly, all contracting States unanimously decided that a new developed programme, based on the DGCA/97 action plan, should be applied systematically and regularly to all States concerned as of 1 January 1999.[4]

As a result, a universal safety oversight audit programme has been established, comprising regular, mandatory, systematic and harmonized safety audits to be carried out by ICAO on all contracting States.

The audits are designed to determine the status of implementation by contracting States of the critical elements of a safety oversight system as well as relevant ICAO SARPs, associated procedures, guidance material and safety-related practices.

If a State has made no significant progress following a reassessment, the results should be published. The name of the programme has been changed to ICAO Universal Aviation Safety Oversight Programme.[5]

Despite the initial reluctance of the contracting States to be subjected to safety audits, together with the inevitable associated financial troubles, eventually safety audits were increasingly seen as an essential tool to enhance aviation safety.

A large list of serious shortcomings emerged from the audits. The results of a 1997 ICAO overall aviation safety survey, presented at an ICAO conference in November 1997, revealed that a number of contracting States did not have or incorrectly applied the ICAO SARPs. The main causes were carelessness, lack of capacity or not really knowing how to deal with the application methods.

The States concerned have unanimously backed a proposal to strengthen the ICAO programme regarding safety oversight. It was recognized that by expanding the ICAO programme to a universal, mandatory and transparent instrument, it would be, in view of global information exchange, preferable to a rather obscure voluntary auditing process. A support base for the mandatory auditing/assessment covenant was obtained by a State-specific Memorandum of Understanding, an instrument of agreement for the conduct of the audit and follow-up actions, based on a generic MoU approved by the ICAO Council and signed by the Secretary General and the appropriate authority of the State to be audited. The result of this initiative has been mentioned before.

---

4 DGCA/97: Conference of Director General of Civil Aviation on Global Strategy of Safety Oversight in 1997. *See*: M. Milde, 'Aviation Safety Oversight: Audits and the Law', XXVI *Ann. Air & Space Law* 2001, p. 165, at 174.

5 ICAO Assembly Resolution A32-11 Establishment of an ICAO USOAP.

A remarkable fact was that the modified programme from that time on would mean distinct mandatory audits for all contracting States, which by Weber, former Head of the ICAO Legal Office, should not be seen as an infringement on the sovereignty of the contracting States. States that fail to sign and submit a copy of the MoU within the prescribed period shall be reported to the Council. All contracting States will be informed of the State's refusal to sign and accept an ICAO safety oversight audit. However, nearly all contracting States participated without significant aversion in the audit procedure with a view to improving State safety oversight and worldwide information exchange, in spite of the strong sovereignty principle.

> Sovereignty is always a sensitive issue. In this instance, however, the aviation community supports the concept of mandatory audits as it recognizes that the need for improvement in safety oversight is global. The problem is not limited to a region or a nation. A system of audits related to the implementation of safety-related SARPs would not represent the infringement on State sovereignty, bearing in mind the fundamental importance attached to safety by the Chicago Convention, and provided that the system itself is agreed upon in advance by the member States.[6]

During the 33rd Session of the ICAO Assembly, from 25 September till 5 October 2001 in Montreal, the results of all conducted audits throughout the world were presented. These results revealed that the implementation rate was based on questionnaires of the protocols. The relevance of standards and audits demonstrates the amazing correlation with aircraft accident rates. States that already have been audited voluntarily, showed virtually no deficiencies following the mandatory assessments used. However, it seems that many of the States audited have not even established basic aviation legislation, or they have existing but rather outdated legislation. As a consequence these States do not take fundamental aspects into consideration, for example to provide the need for adequate enforcement powers. In addition, it was found that in many contracting States, the entities, which should regulate and monitor aviation safety, indeed have the disposal of the necessary independence and authority to properly fulfil their regulatory duties.

The ICAO USOAP audit team also examined the Dutch Civil Aviation Department and the Department of Civil Aviation of the Netherlands Antilles. In August of 2001, the final results were handed over by ICAO. The report contained 17 findings and recommendations regarding the Netherlands and 18 for the Antilles.

6 A. Kotaite, 'Universal Audit Programme Conducted under Auspices of ICAO Would Offer Many Benefits', *ICAO Journal*, Vol 53, No 4, 1998, at 6.

In consultation with ICAO and in accordance with the adopted procedure, the Netherlands and the Netherlands Antilles have drafted an action plan indicating how to deal with the recommendations.

6.2 ICAO Safety Initiatives

Enhancing aviation safety will be one of the primary objectives of ICAO in the coming years.[7] The Air Navigation Commission proposed in 1997 the Global Aviation Safety Plan (GASP) to the ICAO Council, recognizing the coordinating role of ICAO with respect to the many safety initiatives emerging on a global scale, all with the common aim of reducing the number and rate of aircraft accidents. In 1998 during the 32nd Assembly Session, Resolution A32-15 concerning the GASP was adopted.

The ICAO Assembly unanimously supported GASP being an essential and effective tool to enhance aviation safety on a global scale. In 2001, Resolution A33-16, containing an updated version of the GASP, was adopted.[8]

The objectives of the GASP are to:

a) reduce the number of accidents and fatalities worldwide irrespective of the volumes of air traffic; and
b) achieve a significant decrease in accident rates, particularly in regions where these remain high.

In addressing these objectives, GASP concentrates on three fundamental aspects of a safety management system, as follows:

> 1st Fundamental reviewing the causal factors of aircraft accidents worldwide in order to identify specific safety issues which must be addressed in order to reduce accident numbers and rates. Attention will be given to the reasons for regional variations in accident rates;
> 2nd Fundamental keeping abreast of the activities of existing safety groups in order to identify safety issues which have global perspectives. In doing this, GASP focuses on those safety initiatives most likely to reduce accident numbers and rates; and
> 3rd Fundamental promoting safety awareness worldwide by facilitating the effective sharing and use of aviation safety data and information.[9]

---

7 L. Weber, 'Convention on International Civil Aviation – 60 Years', 53 *Zeitschrift für Luft- und Weltraumrecht* 2004, at 289-311.

8 In October 2004 the 35th Session of the Assembly reviewed a progress report on GASP that included an updated version of the GASP document.

9 The ICAO Global Aviation Safety Plan (2004 Edition), (1) Objectives and (2) Fundamentals.

The GASP serves all parties involved in aviation safety, including acting as a planning and tracking tool to monitor progress in the relevant areas of activity, and will focus on the safety-related activities within ICAO and those safety initiatives, either planned or in progress, that offer the best safety dividend in terms of reducing the aircraft accident numbers and rates worldwide.

From early 2004 it was decided to expand the mandatory audits on contracting States to Air Traffic Services (Annex 11) and Aerodromes (Annex 14).

Furthermore, the Secretary-General was commissioned to investigate the possibility of including Annex 13 (Aircraft Accident and Incident Investigation) in GASP.

Moreover, the suggestion was made to gradually record USOAP in the regular budget. The Assembly has drawn special attention to the implementation of action plans resulting from audits, especially to developing countries. Support was given to the introduction of a quality assurance mechanism in order to qualitatively ensure technical assistance and to embed it as an independent unit in the ICAO organization. In March 2004, the Council decided to facilitate the evolution of USOAP from an Annex-to-Annex approach method to a Comprehensive Systems Approach, which would cover all safety-related Annexes and focus on the overall capability of States for safety oversight.

In a European review on the value of the ICAO USOAP and its expansion, 38 ECAC member States, including 15 European Community member States, presented, during the 33rd Assembly Session in 2001, a paper reaffirming European support for the continuation of the existing USOAP audits. The member States consider that the continuation of a regular programme of follow-up audits of Annex 1, 6 and 8 is of the highest priority to the achievement of ICAO's objectives.

The lessons learnt from the existing USOAP about legal and regulatory frameworks should have general application to other aviation safety fields. However, the specific information about compliance with SARPs in other areas is also important for States since they need to assure themselves about the safety of their aircraft and citizens worldwide. In principle, therefore, according to ECAC, it is desirable for the programme to move into areas covered by other Annexes to the Chicago Convention.

ECAC member States agreed in 2001 that there should be a step-by-step expansion of USOAP as soon as practicable, provided financial resources were available. In assessing priorities among the Annexes to be addressed in an expanded programme, the importance of the areas covered in safety terms should be the primary consideration. On this basis, the area covered by Annex 11 (Air Traffic Services) is arguably the most important, followed by Annex 14 (Aerodromes) and then Annex 13 (Aircraft Accident and Incident Investigation). Although ECAC member States consider that full audits of Annex 13 may be of a lower priority than audits of Annexes 11 and 14, they note that there have been problems with the implementation of Annex 13 in some contracting States.

Finally, in their 2001 statement, ECAC member States reaffirmed their belief that real safety benefits will result only from the collective ability of contracting States to correct deficiencies whenever they occur. Collective actions should take advantage of the range of existing support mechanisms, as well as to develop new ones as necessary.

At the 35th Session of the ICAO Assembly, according to Assembly Resolution A35-6, it was decided to continue USOAP, but to change the Annex-by-Annex approach to a Comprehensive Systems Approach as of 1 January 2005, which would cover all safety-related Annexes (except Annexes 9 and 17, which are subject to the Universal Security Audit Programme) and focus on the overall capability of States for safety oversight. In Europe EUROCONTROL established in 2002 an audit programme (ESARR Implementation Monitoring and Support ESIMS) to assess States' safety oversight arrangements and capabilities, covering all applicable regulations related to the implementation of ATM safety oversight by a State. In view of the close synergies between USOAP and ESIMS, the programmes operate as from 2005 in close, formal cooperation.

Until 2006 almost all contracting States have been visited for Annex-by-Annex and Comprehensive Systems Approach safety audits by USOAP, in such a way that we can emphatically speak of a global safety audit experience.[10] The audits focus on the topics contained in the Annexes 1, 6 and 8 with links to the other safety-related Annexes, such as 16 (Environmental Protection) and 18 (Safe Transport of Dangerous Goods by Air). The sincere but also coercive desire of contracting States to display and ascertain transparency of own and others' national affairs regarding aviation safety oversight is expressed by the "status of release consent provided by States of USOAP audit report information".[11]

A source of concern remains the financial fundraising with respect to technical assistance projects, particularly for developing countries. Several years ago a number of States suggested the creation of a plan called International Financial Facility for Aviation Safety (IFFAS) regarding these projects.[12]

> The origin of an International Financial Facility for Aviation Safety dates back to 1995 when the 31st Session of the ICAO Assembly considered a proposal from a group of States for study of the need, appropriateness and usefulness of establishing an international aeronautical monetary fund to finance investments in airport and air navigation services infrastructure under conditions that would be more flexible and less onerous than those usually available in financial markets. Preliminary works on the proposal was carried out in the 1995-1998 triennium and the concept was endorsed by the Directors General of Civil Aviation Conference on a Global Strategy for Safety Oversight (1997)

10 ICAO Working Paper DGCA/06-WP/3 of 9 January 2006. The Status of Safety Oversight. Only Sierra Leone, Burundi, Afghanistan, Iraq, Liberia, Solomon Islands and Somalia fell outside the ICAO USOAP audits for security reasons.

11 <www.icao.int/fsix/safety.cfm> (21 December 2011).

12 *See also* ICAO Doc 9848 at I-61 (A35-8 International Financial Facility for Aviation Safety): IFFAS will provide financial support towards achieving the objectives of improving aviation safety, through the implementation of the necessary corrective measures identified in the ICAO USOAP.

> and the World-wide CNS/ATM Systems Implementation Conference (1998). Subsequently, in 1998 the 32nd Session of the Assembly noted the progress on this project and endorsed plans for further work on a plan.
> The 33rd Session of the Assembly, by Resolution A33-10 Establishment on an International Financial Facility for Aviation Safety noted work carried out during the triennium, endorsed the concept of the IFFAS, *inter alia*, and requested the Council to pursue the establishment of an IFFAS as a matter of priority early in the 2002-2004 triennium, having due regard to the applicable laws of contracting States. This request carries with it the formulation of appropriate management, administrative and legal strategies towards the initial implementation of the IFFAS within the 2002-2004 triennium.[13]

IFFAS is a fund based on voluntary contributions of the contracting States and other stakeholders in the civil aviation industry. It is considered to be independent of ICAO's regular budget program and is designed to assist states to correct safety-related deficiencies identified through ICAO USOAP audits. Initially there was considerable resistance because of the lack of clarity on how to financially feed this fund. During the 33rd Session of the Assembly an agreement was reached on the issue of (voluntary) contributions to this fund and the strategy regarding development and implementation of this instrument.

IFFAS commenced implementation of its funding mechanism in 2004. With limited resources, priority is given to projects proposed by least developed countries.

Regarding aviation safety in upcoming States on the African continent, ICAO has set forth principles and recommendations in the Comprehensive Regional Implementation Plan for Aviation Safety (AFI Plan, adopted by the 36th Session of the ICAO Assembly in September 2007). To action the key priorities established in the AFI Plan, the AFI Comprehensive Implementation Programme was established in early 2008. The broader objective of the AFI Plan is to enhance aviation safety through increased cooperative efforts by States, in conjunction with global industry stakeholders and under the leadership of ICAO, in order to effectively address deficiencies recently identified through the USOAP. To enhance cooperation and bring States, service providers and operators, together at the political and technical levels to deal with aviation safety aspects, the Africa, Indian Ocean Regional Air Navigation (AFI-RAN) meeting was organized in Durban, South Africa, in November 2008. The objective of this meeting was to offer an opportunity to AFI States to strengthen region-wide commitments to resolve deficiencies, address critical safety issues and develop a comprehensive work programme for the implementation of improved air navigation facilities and services.

The 36th Session of the ICAO Assembly urged contracting States of the AFI region to commit themselves to achieving the goals and objectives of the AFI Plan and that the

13 *Source*: ICAO Air Transport Bureau, Secretary IFFAS.

ICAO GASP be used as the framework for the development of regional and sub-regional implementation plans, ensuring harmonization and coordination of efforts.

On the basis of analyzes of USOAP audits and ICAO accident/incident data reporting (ADREP), the meeting urged AFI States to make it a high priority to implement all eight Critical Elements of a safety oversight system with special emphasis on those that have a strong relationship with accident rates, namely: Critical Element 4 (Qualification and training of technical personnel); Critical Element 8 (Resolution of safety concerns); Critical Element 6 (Licensing and certification); Critical Element 3 (State civil aviation system and oversight functions); and Critical Element 7 (Surveillance obligations). Furthermore, the meeting urged African States, with the assistance of ACIP, to acquire the capability for effective and sustainable safety oversight systems in coordination with COSCAPs and Regional Safety Oversight Organizations (RSOOs), apart from prompting various countries to establish potential accident investigation organizations and safety oversight bodies as well as improving the safety and efficiency of the air navigation systems on the African continent by adopting the performance-based framework for air navigation system planning and harmonizing the work of the AFI region and other ICAO regions.[14]

One of the basic tenets of the ICAO USOAP is transparency. It means that information gained in the course of the ICAO oversight audits will be shared with all contracting States. In this context, ICAO attempts to achieve this determined goal by disseminating summary reports of its audits to all other contracting States to give them the opportunity to form an opinion of the safety oversight being provided by the civil aviation authority of the audited State.

The needs of audited States are equally important in this regard, *i.e.* their deficiencies will not be effectively susceptible to correction if they are not carefully diagnosed and then adequately depicted in sufficient detail in ICAO audit reports. It has been noted that many civil aviation authorities have found that such reports provide effective leverage with higher-level governmental authorities to create or enhance political will and, hence, help them to secure adequate resources in order to comply with ICAO safety oversight provisions.

6.3 CONCLUDING REMARKS

In principle, aviation safety oversight is a national responsibility. Aviation safety oversight may apply to State safety oversight the process of ensuring that air carriers, aircraft, aircrew, maintenance staff, manufacturers and a host of others who are engaged in civil aviation perform their functions in a safe and responsible way to ensure that applicable safety regulatory requirements are met. The mainstream of daily operation activities regarding

14 ICAO Regional Report Africa 2009 by ICAO Regional Officer B.M. Sekwati.

safety oversight at the civil aviation authority level, such as comprehensive monitoring of compliance with international standards, quality control, issuing certificates and implementing regulation-based executory decisions and policy measures, will be evaluated by means of worldwide standardized safety oversight audits based on programmes created and conducted by organizations such as ICAO, which recognizes national safety oversight responsibilities within States' obligations under the Chicago Convention.

ICAO has published ICAO Doc 9734 Part A – The Establishment and Management of a State's Safety Oversight System, and Part B – The Establishment and Management of a Regional Safety Oversight System, giving comprehensive guidance information to support the development of safety oversight. In addition, ICAO has established its USOAP, with the aim of verifying, through formal audits, States' safety oversight arrangements and capabilities.

Regular, systematic, harmonized and mandatory audits cover all applicable regulations related to safety oversight by a State, as well as the procedures, the resourcing of functions and other implementing arrangements that are needed to implement the system's critical elements as defined in ICAO Doc 9734 A:

- Primary Aviation Legislation,
- Specific Operating Regulations,
- State Aviation System and Safety Oversight Functions,
- Qualified Technical Personnel and Training,
- Technical Guidance, Tools and Provision of Safety Critical Information,
- Licensing, Certification, Authorization and Approval Obligations,
- Surveillance Obligations, and
- Resolution of Safety Concerns.

The audits are designed to determine the status of the States' implementation of the critical elements and the implementation of relevant ICAO SARPs and associated procedures, laid down mainly in Annexes 1, 6 and 8 to the Chicago Convention as well as guidance material and safety-related practices.

However, very recently, an unexpected negative turn regarding State safety oversight strengthening, particularly in the European and North Atlantic region, became ruthlessly evident from USOAP audits.

> States are chronically under-resourcing the aviation safety oversight functions of their national aviation authorities, particularly in the European and North Atlantic region. ICAO discovered this lack of resources at national levels through USOAP audits. The cause appears to be government austerity measures, and ICAO fears the under-resourcing could become even more severe, because many audits that revealed the deficits were evident before reaction

> to the global economic crisis started to kick in. Reports from USOAP audits indicate lack of effective safety procedures implementation globally in 42% of States, and in 40% of European and North Atlantic countries, with this directly attributable to lack of financial resources.[15]

The greatest problem in this downturn of safety procedures implementation, affecting safety oversight, is the lack of human resources, owing to a combination of posts not being filled or posts not being properly staffed, mostly by people unqualified for key positions. The centre of the issues, austerity measures taken to downsize State institutions, is often at high political levels and demonstrates a lack of insight and responsibility regarding the adverse consequences.

Highly developed Western contracting States, in particular, are setting an extremely bad example. According to Nicolas Rallo, ICAO Regional Office Paris Safety Officer, ICAO fears that, as safety management systems become mandatory in the European and North Atlantic region and elsewhere, governments may start to delegate industry monitoring and oversight to the industry itself, and this would amount to far-reaching deregulation. This disturbing downward evolving trend will most probably further reduce regulatory oversight resources, and at the same time could introduce unwanted erosion of aviation safety.

15 D. Learmount, 'ICAO Reveals Chronic Under-Resourcing of Safety Oversight', *Flight International* 5 March 2012.

7 SAFA Inspection Methodology

7.1 References

In each SAFA participating State, aircraft (third-country for EU member States or foreign for non-EU ECAC States) can be subject to a ramp inspection, which is normally carried out at random.[1] The civil aviation authorities of these participating States are in fact only able to inspect aircraft of air carriers that operate flights to and from Community airports, and in view of the rather random nature of such inspections, it is not possible to check all aircraft that land at each Community airport. These ramp inspections focus on aircraft documents (*e.g.* Certificate of Airworthiness) and manuals, flight crew licences, the apparent interior and exterior condition of the aircraft and the on-board presence, location and condition of mandatory cabin safety equipment as well as the stowage and the type of on-board cargo.

The references for a SAFA inspection are contained in the European Basic Regulation and its implementing rules as well as the provisions outlined in the Chicago Convention and in the minimum standards of the ICAO Annexes that are commonly applicable on a global scale to all inspected aircraft. In particular, the standards of Annex 1 (Personnel Licensing), Annex 6 (Operation of Aircraft) and Annex 8 (Airworthiness of Aircraft) with links to Annex 16 (Environmental Protection), Annex 17 (Security – Safeguarding International Civil Aviation against Acts of Unlawful Interference) and Annex 18 (The Safe Transport of Dangerous Goods by Air).[2] In addition, when inspecting the technical status of an aircraft, it should be checked against the aircraft type certification and the standards set by the aircraft manufacturer. Furthermore, compliance with national standards that are declared applicable to all air operators flying to that particular State of Inspection may be checked.[3] If non-compliance with a national standard is detected, this standard should be explicitly stated in the SAFA Ramp Inspection Report.

---

1 EASA has signed working arrangements with 15 non-EU ECAC States involving them in the SAFA programme. The EU member States as partners in the SAFA programme and the non-EU ECAC States are therefore together called the SAFA participating States. The 42 participating States (2011) engaged in the EC SAFA programme are: Albania, Armenia, Austria, Azerbaijan, Belgium, Bosnia and Herzegovina, Bulgaria, Croatia, Cyprus, Czech Republic, Denmark, Estonia, Finland, France, Germany, Greece, Hungary, Iceland, Ireland, Italy, Latvia, Lithuania, Luxembourg, Malta, Moldova, Monaco, Netherlands, Norway, Poland, Portugal, Republic of Georgia, Romania, Serbia and Montenegro, Slovakia, Slovenia, Spain, Sweden, Switzerland, the former Yugoslav Republic of Macedonia, Turkey, Ukraine, United Kingdom.

2 ICAO Annex 16, Vol I, 5th edn, July 2008. ICAO Annex 18, 3rd edn, July 2001, Amendment 9, November 2008.

3 EASA Standardizations & Approvals Directorate SAFA Coordination Section. SAFA Ramp Inspections Guidance Material Version 1.0 July 2009, at 9.

A SAFA ramp inspection will be carried out according to a specific procedure that is common to all participating States.

EASA, as a coordinating entity regarding civil aviation in the European region and beyond, is playing a decisive role in the SAFA programme, although the European Commission carries overall responsibility and has legislative powers.

In accordance with Article 2 of Commission Regulation (EC) No 768/2006,[4] as well as Article 9 (aircraft used by a third-country operator into, within or out of the Community) and Article 23 (Third-country operators), of the Basic Regulation, EASA has been given a number of new tasks and tools with regard to the collection and exchange of information on the safety of aircraft using Community airports and the exclusive management of the information system, and regarding any other relevant information with respect to operators of aircraft registered in a third-country or registered in a member State that has delegated their appropriate regulatory safety oversight to a third country and used by a third-country operator into, within or out of the Community.[5]

Non-EU ECAC member States engaged in the EC SAFA programme and EU member States, together represented in the ESSG (European SAFA Steering expert Group), are advising the European Commission on some specific SAFA-related subjects:

1. Assisting the European Commission in a consultative manner in relation to proposals from EASA prior to their submission to the Air Safety Committee, regarding the overall safety oversight policy of the EC SAFA programme, possible actions to be taken in specific safety cases and recommendations for the further development of the SAFA programme, including the amendment of SAFA procedures;
2. Communicating to the Commission feedback upon the national implementation of the SAFA programme in their State;
3. Reviewing and reacting upon the SAFA database analysis that EASA is required to conduct on a frequent basis.[6]

The central SAFA database, transferred in 2006 to EASA, provides the possibility for enhanced exchange and sharing of information between the States participating in the SAFA programme. Information from all conducted ramp inspections is collected, protected, stored, analyzed and disseminated for the benefit of each participating State, in particular,

---

4 Commission Regulation (EC) No 768/2006 of 19 May 2006 implementing Directive 2004/36/EC of the European Parliament and of the Council as regards the collection and exchange of information on the safety of aircraft using Community airports and the management of the information system, *Official Journal of the European Union L* 134/16, 20 May 2006.

5 Article 1.2 of Regulation (EC) No 216/2008 states: This Regulation shall not apply when products, parts, appliances, personnel and organizations referred to in paragraph 1 of this Article are engaged in military, customs, police, or similar services (State aircraft). The member States shall undertake to ensure that such services have due regard as far as practicable to the objectives of this Regulation.

6 <www.easa.europa.eu/approvals-and-standardisation/safety-assessment-of-foreign-aircraft-SAFA.php>.

and for other parties concerned, in general, thus contributing to a more balanced view concerning a specific aircraft, aircraft type or operator. However, limited time inherent to an – on-time performance – aircraft turn-around (without delay) and consequently a superficial (SAFA) ramp inspection, consisting of only 54 check items, will not yield a full picture regarding compliance with applicable and essential safety requirements of that particular aircraft or operator.[7] Of the checklist items, 24 relate to operational requirements (A-items) to be checked in the cockpit, 14 items address safety and cabin issues (B-items), 12 items concern the aircraft (technical) condition (C-items), and 3 items are related to the inspection of cargo (dangerous goods) and the cargo compartment(s) (D-items). Certain aspects concerning more complex and profound issues such as the full airworthiness status of an aircraft, crew resource management training, compliance with maintenance schedules and operational flight procedures remain difficult, if not impossible, to fully assess during a single ramp inspection.

A comprehensive safety assessment can be achieved only through the continuous oversight conducted by the responsible civil aviation authority of the State of the Operator or the State of Registry in cooperation with inspecting State authorities to resolve and eliminate specific safety-related problems.[8] Nevertheless, information gained by SAFA inspections is certainly useful:

- Primarily as a pre-emptive tool helping to identify potential negative safety trends whereby a numerous and/or recurring number of findings concerning a particular operator will be a very good indicator of potential structural weaknesses both with regard to the quality control management of that operator as well as the level of safety oversight exercised by the responsible national civil aviation authorities of the State where that operator is certified; similar negative trends may also be identified concerning specific aircraft types.
- More directly, SAFA inspections may contribute in real-time to the safe operation of the particular aircraft which has just been inspected, prompting the inspection authorities to ensure that corrective actions are taken immediately, prior to any further operations being conducted by that aircraft.[9]

In addition, the SAFA programme, already an established and consolidated overall European safety-monitoring tool, is reinforced by the European Commission through maximizing the level of harmonization of procedures for the conduct and reporting of ramp

7 The 54 check items are mentioned in Annex II to Commission Directive 2008/49/EC of 16 April 2008 amending Annex II to Directive 2004/36/EC of the European Parliament and of the Council regarding the criteria for the conduct of ramp inspections on aircraft using Community airports, *Official Journal of the European Union L* 109/17, 19 April 2008.

8 The responsibility of a State Civil Aviation Authority extends to regulatory functions such as certification and licensing and safety oversight functions to ensure compliance.

9 The EC SAFA Programme, Past, Present and Future (<ec.europa.eu/transport/air/safety/safa_en.htm>).

inspections in all participating States (SAFA Procedures Manual), for a common set of criteria for the training and qualification of SAFA inspectors, including the issuing of detailed guidance material regarding uniform and standardized rules.

The qualification criteria and training syllabi, as defined and developed by EASA, focus on eligibility, training, recurrent courses, using confidential data from the SAFA database, and experience of inspectors.[10]

A rather new perspective feature developed in the context of the SAFA programme and based on exchange of information is the alerting function. The SAFA alarm function became fully operational on 15 May 2006.

The EUROCONTROL CFMU (Central Flow Management Unit) is alerting ECAC member States to flight plans to and from ECAC airports pertaining to aircraft or operators that have been subject to a prior SAFA alarm message, notifying that these aircraft or operators are subject to an operating restriction. Member States should be allowed in exceptional cases to take unilateral measures when confronted with unforeseen safety problems, for example to instantly impose an operating ban, although in the interest of a common Community approach this power should be of a restrictive nature.[11]

To minimize or even eliminate operating bans, the SAFA programme will maximize assistance to future EU member States and third countries that want to develop privileged links with the Community and apply its safety standards, sharing resources and enhance national civil aviation safety oversight activities.

EASA will progressively assist the European Commission and EU member States with their cooperation, technical working arrangements and additional assistance activities with third countries that are running short of human (expertise) and financial resources.

7.2 Ramp Inspection

Ramp inspections in States participating in the EC SAFA programme are carried out at random or in a more focused manner if there is reason for alleged non-compliance with ICAO minimum requirements by a particular third-country aircraft or operator. The common non-discriminatory inspection procedure takes place during a regular scheduled or non-scheduled aircraft turnaround at an airport open to international air traffic located in a participating State. In order not to cause any undue delay, inspection time may be insufficient to go through the full checklist of 54 items, restricting the inspection team to less than maximum coverage. In that case, inspectors should concentrate on those items that, based on experience, preparation and (non-discriminatory) expectations, are likely

10 *See* Commission Directive 2008/49/EC.

11 Regulation (EC) No 2111/2005.

to be more safety critical. However, if an inspection reveals issues that may seriously affect flight safety or discloses more complex safety matters requiring additional research time for clarification, a delay may be justified.

It has to be emphasized that a single SAFA ramp inspection will give a limited, approximate impression regarding the safety level of the aircraft or the operator in question and could not be considered a substitute for eligible aviation safety oversight.

The checks are carried out according to a procedure that is common to all participating States and should be performed in a tactful, effective and least inconvenient way, especially with respect to flight and cabin crew, passengers, handling agents and other personnel involved in ground handling activities. Attention should be given to the fact that an inspection should not jeopardize the total rest time of the flight crew during the relatively short turnaround phase (aircraft on blocks at the ramp).

Before the actual inspection is conducted, but after the aircraft/operator has been selected and relevant information, among other things, derived from the SAFA database preparations of inspections module, is collected, the next step will be the preparation phase for the planned inspection, consisting of:

- Last update of the operating schedule for the selected operator should be obtained from airports, operators or ground handling officials;
- Distribute the tasks between the SAFA inspectors involved, especially in the case of a limited inspection time and/or the size and the complexity of the aircraft;
- Select the checklist items to focus on and take into account the data from previous inspections collected from the database;
- There should be cooperation with security, ground and all other officials involved in airport activities, to enable the SAFA team to reach the aircraft to be inspected;
- For planning purposes, EUROCONTROL has a special module in which you will be able to find relevant flight information concerning any targeted operators. The application form to request access to the CFMU system can be found on their website.

A SAFA inspection team consists of at least two inspectors. Inspectors must have received training in their relevant field of expertise before being qualified.[12] Qualified inspectors are entitled to perform a SAFA ramp inspection and search the aircraft in question according to and under Article 16 (Search of Aircraft) of the Chicago Convention:

> The appropriate authorities of each of the contracting States shall have the right, without unreasonable delay, to search aircraft of the other contracting States on landing and departure, and to inspect the certificates and other documents prescribed by this Convention.

12 *See* EASA Guidance Material of the Qualification of Inspectors.

A participating State or the competent authority is responsible for establishing a schedule for the performance of SAFA ramp inspections. EU member State authorities shall prioritize these ramp inspections on the subjects included in the list referred to in Article 3 of Commission Regulation (EC) No 351/2008.[13]

The schedule for the performance of SAFA inspections may consist of:

a. Long term planned inspections. Operators performing scheduled operations could be selected on a long term basis for inspection since their schedule is known to the competent authority. In the absence of any specific suspicion of non-compliance with the international safety standards those inspections could be seen as planned – spot-check inspections. Long term selection should be done considering the following:
   - repeated inspections should be avoided on those operators on which previous inspections have not revealed safety deficiencies;
   - the selection would enable the widest possible sampling rate of the operator population flying into their territory;
   - no discrimination, based on the nationality of the operator, the type of operation or type of aircraft.
b. Short term/ad-hoc planning. Ad-hoc planning could be used when information is received on suspected or verified non-compliance with the international safety standards.[14] Such information might be originating from, but not limited to:
   - the list for prioritized ramp inspections under the EC SAFA programme;
   - previous inspections done by the member State;
   - the SAFA database (inspections from other member States);

13 Commission Regulation (EC) No 351/2008 of 16 April 2008 implementing Directive 2004/36/EC of the European Parliament and of the Council as regards the prioritization of ramp inspections on aircraft using Community airports, *Official Journal of the European Union L* 109/7, 19 April 2008. Article 3.1 states: A list of subjects referred to in Article 2 shall be communicated electronically by EASA to the member States at least every four months. Article 2 states: Without prejudice to Article 8(3) of Directive 2004/36/EC, member States shall prioritize their ramp inspections on the following subjects landing at any of their airports open to international air traffic:
   1. Subjects identified as posing a potential safety threat on the basis of the regular analyses conducted by EASA.
   2. Subjects identified on the basis of an opinion expressed by the Air Safety Committee within the context of the implementation of Regulation (EC) No 2111/2005 that further verification of effective compliance with relevant safety standards through systematic ramp inspections on those subjects is necessary. This may include subjects which have been withdrawn from the list of air carriers subject to an operating ban within the Community established by Regulation (EC) No 2111/2005 (the Community list).
   3. Subjects identified on the basis of information obtained by the Commission from the member States or EASA pursuant to Article 4(3) of Regulation (EC) No 2111/2005.
   4. Aircraft operated into the Community by operators included in Annex B of the Community list.
   5. Aircraft operated by other operators certified in the same State as any operator featuring concurrently on the Community list.

14 Aircraft being suspected or verified on non-compliance with international safety standards shall be subjected to an obligatory ramp inspection. The legal obligation to conduct inspections: *see* Article 4 Ramp inspection, Directive 2004/36/EC.

- the department granting traffic rights (*e.g.* when any new type of operations or new type of aircraft being introduced);
- passenger complaints;
- Air Navigation Service Providers (reports that an aircraft has performed abnormal manoeuvres which give rise to serious safety concerns since entering the airspace of the member State);
- 'whistleblowers' (ground handling or maintenance personnel) regarding poor maintenance, obvious damage or defects, incorrect loading, etc;
- evidence that the State in which the aircraft is registered may not be exercising proper safety oversight; or
- concerns about the operator of the aircraft which have arisen from occurrence reporting information.

c. Spot checks.[15] A certain amount of the total number of scheduled inspections may be reserved for performing inspections on a random check basis. These inspections typically are not initiated by any suspected or verified non-compliance; selection is very often done shortly before the arrival of the aircraft or even when the aircraft is on blocks. The considerations as mentioned in Part a. should be taken into account as far as practical.[16]

The obligations set out in Directive 2004/36/EC (Annex II to this Directive regarding the criteria for the conduct of ramp inspections on aircraft using Community airports has been amended by the Commission Directive 2008/49/EC) pertain to EU member States only.

However, EASA has signed working arrangements with the 15 non-EU ECAC States involving them in the SAFA programme and consequently linking them to these obligations.

The results of the SAFA checks are combined in reports that also follow a common format. If the State of Inspection finds significant irregularities, the operator and the appropriate aviation authority of the State of Oversight, generally the State of the Operator, respectively, but not necessarily the State of Registry or the State of Licensing (in case of licence irregularities or forgery), are contacted in order to come to the imposition and execution of corrective measures at the spot. The competent authority is responsible for the oversight of the operator, the aircraft and/or the personnel licensing of the flight crew, in other words responsible for the oversight in the relevant fields, which could be assigned to different States, owing to the possibility of transfer under Article 83 bis of the Chicago Convention.

---

15 Member States may carry out ramp inspections in accordance with a spot check procedure in the absence of any particular suspicion, provided that they comply with Community and international law (Directive 2004/36/EC under (7)).

16 EASA Standardisations & Approvals Directorate SAFA Coordination Section. SAFA Ramp Inspections Guidance Material, Version 1.0, at 7.

The corrective measures might not only be applicable to the aircraft in question, but also to other aircraft when there is serious doubt about the irregularities being once-only or of a structural, recurring nature.

However, if the State of Inspection reveals major findings directly affecting the safety of the aircraft, crew and passengers or even potential third parties, the authority of that State may request immediate corrective actions before subsequent departure. If more extensive repair time is needed or no adequate repair facility is present, the authority of the State of Inspection may decide, in coordination with the State of the Operator, the State of Registry or the State of Licensing of the aircraft and its crew in question, to authorize a non-revenue positioning flight to a suitable airport and to dictate the necessary stipulations under which such a ferry flight must be carried out.

7.3 Findings and Their Categorization

During an aircraft pre-flight inspection, covering all critical technical and operational items, any existing defects should be detected by line-maintenance personnel or the operating flight crew at that time. It is the operator's responsibility to address these (obvious) defects before commencing the flight. If the operator did not detect the defect but should have identified it because it was obviously visible to the SAFA inspector during the SAFA ramp inspection, that inspector should inform the crew on the non-identified defect(s). On the other hand, if the operator detected the defect but did not properly report and/or assess it, the operator should be required to assess the defect, in both instances irrespective of whether the defect was within or outside dispatch limits. Not noticing or ignoring any obvious defects by the operator may give rise to official reporting, more frequent inspections or even grounding. If during an inspection it is established that a certain situation is not in compliance with the relevant standards, then it is considered to be a finding.

Findings determined during a SAFA inspection are categorized according to the magnitude of the deviation from the safety standards and to the influence or impact on flight safety of the aircraft not being compliant.[17] However, non-compliance with an internationally agreed safety standard does not necessarily mean a direct and catastrophic threat to the safety of the aircraft and its occupants.

The primary goal of categorizing the SAFA inspection findings is to classify the compliance with a standard and the seriousness of non-compliance with this standard. For each inspection item, three categories of possible deviations from the standards and their actual perceived influence on flight safety have been defined: a category 1 finding is considered

17 *See also*: Comment Response Document to Notice of Proposed Amendment 2008-22B and 2009-02D, AR.GEN.440 Categorization of findings.

to have a minor influence on flight safety (for example: copy of the Operational Flight Plan not signed by the aircraft commander).

A category 2 finding may have a significant influence (for example: discrepancies between computerized and actual fuel data), and a category 3 finding may have a major impact on flight safety (for example: miscalculation of the aircraft mass and balance).[18]

Category 1 findings are notified to the aircraft commander by means of a verbal debriefing and the delivery of the proof of inspection as a Class 1 action. Such action shall be taken after each inspection, regardless of whether findings have been identified or not. Category 2 findings are additionally communicated to the operator and in writing to the Civil Aviation Authority of the State of the Operator or the State of Registry as a Class 2 action, and shall contain request for evidence of corrective actions taken by the operator concerned and a request for confirmation that the Civil Aviation Authority is content with the corrective actions taken by the operator. Findings identified as category 3 could also be reason for Class 2 actions.

Category 3 findings and subsequent actions, in addition to the actions of a category 2 finding, may include restrictions on the flight operation (Class 3a), corrective actions before the next flight leg (Class 3b), detention and grounding of the aircraft by the inspecting national Civil Aviation Authority (Class 3c) or even revocation of the operator's entry permit, and, worst of all, the imposition of an immediate operating ban under the applicable provisions of national and Community law (Class 3d).[19]

If there is reason to believe that a potential safety threat is at issue, the SAFA national coordinator shall inform, without delay, the European Commission and the EU member States.[20]

The EASA Notice of Proposed Amendment No 2009-02D of 30 January 2009, containing the draft Opinion on the Implementing Rules for Air Operations of Community Operators, the Subparts related to Air Operations of the draft Opinion on the Implementing Rules for Organization Requirements, the Subparts related to Air Operations of the draft Opinion on the Implementing Rules for Authority Requirements and the related

18 Commission Directive 2008/49/EC of 16 April 2008 amending Annex II to Directive 2004/36/EC of the European Parliament and of the Council regarding the criteria for the conduct of ramp inspections on aircraft using Community airports, Annex II Manual of EC ramp inspection procedures – Core elements 5. Categorization of findings. Also 5.2. Instructions on the categorization of findings will be developed and published as detailed guidance material by EASA and amended as necessary in order to reflect relevant scientific and technical progress.

19 An aircraft is grounded in a situation where, following the identification of Category 3 (major) findings, the competent authority performing the ramp inspection is not satisfied that corrective measures will be taken by the aircraft operator to rectify the deficiencies before flight departure, thereby posing an immediate safety hazard to the aircraft and its occupants. In such cases the national civil aviation authority performing the ramp inspection shall ground the aircraft until the hazard is removed and shall immediately inform the competent authorities of the operator concerned and of the State of Registry of the aircraft.

20 Commission Directive 2008/49/EC of 16 April 2008, Annex II.

draft Decisions, explicitly states in Section VI Draft Opinion Part-Authority Requirements (PART-AR) Subpart GEN-General Requirements Section IV Ramp Inspections AR.GEN.440 Categorization of Findings:

> (a) For each inspection item, two categories of possible non-compliances with the applicable requirements are defined as findings.
> Such findings will be categorized as follows:
> 1. A level 1 finding is any significant non-compliance with the requirements or the terms of an approval or certificate which lowers the safety standards and seriously hazards flight safety.
> 2. A level 2 finding is any non-compliance with the requirements or the terms of an approval or certificate which could lower the safety standards and possibly hazards flight safety.

The categories in this case will be a level 1 finding with major influence on safety, and a level 2 finding with significant influence on safety.

7.4 CONCLUDING REMARKS

Immediately following the August 2005 aircraft accidents,[21] ECAC met with the European Commission, EASA, JAA and EUROCONTROL to reaffirm their commitment to maintain and enhance the already high level of safety in civil aviation within Europe. To achieve these goals, the representatives adopted an Action Plan for Safer European Aviation, incorporating several measures already in the policy pipeline within the European Union, such as: Strengthening of the monitoring capability by national and European authorities to ensure full compliance with existing safety rules and procedures, increasing transparency so that passengers know the identity (and safety record) of the air carrier they will be using in advance, establishing European common criteria for imposing European-wide bans on unsafe aircraft/air carriers, simultaneously with strengthening of the system of SAFA inspections of third-country aircraft and adopting measures for assessing foreign air carriers prior to giving them authorization to fly into Europe.[22]

In March 2006, as an important first initiative in response to public concerns over the safety levels of air carriers, the European Union adopted and published the Community

21 Tuninter flight 1153 on 6 August 2005, Helios Airways flight 522 on 14 August 2005 and West Carribean Airways flight 708 on 16 August 2005, killing 297 passengers and crew, including a large number of European citizens.

22 Documents, Working Papers 2006 Ordinary Session (Third part) 26-30 June 2006 Vol IV of the Parliamentary Assembly of the Council of Europe, at 43.

list indicating those air carriers banned from operating within the European Union. This EU blacklist, which is prepared by the Aviation Safety Committee through careful assessment and information of the SAFA programme inspection findings, is based on the principle that carriers banned in one member State are banned in all member States. A fully operational centralized computer-controlled database, governed by EASA, holds confidential (which in this case means not available to the general public) information of ramp inspection reports and supplementary information, such as lists of corrective actions carried out following SAFA inspections. The information is reviewed and analysed by EASA on a regular basis to be able to inform the European Commission and member States of any potential safety hazard identified and notable trends and advise on appropriate follow-up actions.

On behalf and in close cooperation with the European Commission, EASA will develop qualitative criteria with the aim of achieving a more focused approach regarding the SAFA inspection priorities. Categorization of inspection findings is defined in the EASA SAFA Ramp Inspections Guidance Material Version 1.0 of July 2009 (SAFA Categorization) and has a legal basis in the SAFA Directive and Commission Directive 2008/49/EC Annex II Manual of EC SAFA ramp inspection procedures – core elements, paragraph 5 Categorization of findings.[23]

Depending on the level of findings, different actions may take place, from a formal briefing to the aircraft commander to more serious and incantatory corrective actions, *e.g.* operational restrictions, actions before flight, actions at maintenance base, grounding or entry permit revocations.

Many stakeholders are participating in the EC SAFA programme nowadays: the European Commission, EASA, the EU member States and EUROCONTROL which, in turn, provides flight plan information (alerting function) to the other stakeholders regarding aircraft subject to an operational measure such as banning.

The SAFA programme, originally launched in 1996 by ECAC with the support of JAA, was transferred under European Community competence as of 1 January 2007. Upon assuming full responsibility for the management and further development of the programme, the European Commission, assisted by EASA, embarked upon a number of measures intended to reinforce the programme by maximizing the harmonization and uniformity of procedures for the conduct and reporting of ramp inspections throughout Europe as well as related follow-up actions.

23 Commission Directive 2008/49/EC of 16 April 2008 amending Annex II to Directive 2004/36/EC of the European Parliament and of the Council regarding the criteria for the conduct of ramp inspections on aircraft using Community airports, *Official Journal of the European Union L* 109/17, 19 April 2008, Annex II paragraph 5 Categorization of findings.

8 EU Blacklist

8.1 Introduction

In response to several fatal aviation accidents, affecting mostly EU citizens, the European Union was spurred to create a so-called blacklist of air carriers declared unsafe in accordance with international safety standards and, as such, not permitted to land at European airports. The term blacklist is not a very recent phenomenon.[1] The crash of the Birgenair Boeing 757 off the coast of the Dominican Republic in 1996 has led the European Commission to define an aviation safety improvement strategy with the aim, among other things, to be able to assess third-country aircraft concerning their compliance with international safety standards.

This dramatic accident, killing 167 German and 9 Polish tourists and 13 crew members, led the German Government to call for establishing a combined EU blacklist of high-risk air carriers.[2]

Another symptomatic accident that might have accelerated the creation of a combined aviation blacklist of risky air carriers in the European Union occurred in 2004, more than one year before the adoption of Regulation (EC) No 2111/2005 (Community list) by the European Parliament and the Council.

Flash Airlines Flight 604 crashed into the Red Sea on 3 January 2004 shortly after take-off from the international airport of Sharm el-Sheikh, located on the Sinai coastline, most probably caused by spatial disorientation of the captain as pilot flying (PF).[3]

All 148 occupants, mostly French tourists, on board the Boeing 737-300 were killed. The Egyptian and French accident investigators, however, gave divergent, even conflicting explanations regarding the cause of the accident, one exonerating the aircraft captain (commander) and the other blaming him.

1 According to the Henry Holt Encyclopedia, the word 'blacklist' originated with a list that England's King Charles II made of 58 judges and court officers who sentenced his father, Charles I, to death in 1649.

2 In 2005 some EU member States such as France, the United Kingdom, Belgium and the Federal Republic of Switzerland had already issued national blacklists before 2005. *See also* BBC news watch 1 September 2005: Swiss publish aviation blacklist.

3 The U.S. NTSB replied to the final investigation report of the Egyptian authorities and concluded: "The evidence collected during this investigation strongly supports the conclusions that no airplane-related malfunction or failure caused or contributed to the accident and that the accident can be explained by the captain's spatial disorientation and the first officer's failure to assume timely control of the airplane". *See also* Press Release [7 March 2006] – NTSB.

Flash Airlines, an Egyptian air charter company, was discredited in Europe before that tragic accident. In October 2002, Swiss civil aviation authorities performed a routine safety inspection on a Flash Airlines Boeing 737-300 at Zurich international airport. The inspection team disclosed several breaches of safety provisions such as missing pilot oxygen masks, inoperable cockpit instruments, and an insufficient number of oxygen bottles.

After grounding the aircraft for corrective actions the Swiss Federal Office for Civil Aviation banned the air carrier from operating within Swiss airspace. The Swiss Ministry of Transport said it blacklisted the Egyptian air carrier owing to systematic safety problems.

8.2 Causative Events

The sequence of catastrophic events, three consecutive fatal aircraft accidents occurring in 2005, was ultimately the underlying motive for the proposal of the European Commission to establish a list of unsafe air carriers to be banned from operating within Community airspace.

The first accident was Tuninter Airlines flight 1153 from Bari international airport, Italy, to Djerba-Zarais international airport in Tunisia.

The Tunisian ATR-72 ran out of fuel owing to an inadequate amount of fuel loaded for the flight because of an incorrectly installed fuel quantity indicator, replaced by maintenance the night before departure, but essentially different from the type used in the ATR-72.

After both engines had flamed out, the flight crew initiated an emergency descent towards the coast of Capo Gallo, but did not manage to reach the landing runway of the nearby Palermo international airport. The result was a ditching into the sea some 23 nautical miles northeast of Palermo. One crew member and 15 passengers died. The technical investigation by the Italian National Air Safety Board (ANSV *Agenzia Nazionale per la Sicurezza del Volo*) revealed that maintenance personnel in Tunis erroneously installed a fuel quantity indicator designed for the ATR-42 model aircraft, producing erratic indications in the ATR-72 because of different characteristics and mode of operation. The lethal difference was not noticed, neither by maintenance nor the crew during the pre-flight inspection and the first flight leg (ferry flight) from Tunis to Bari nor by the refuelling operator or the flight engineer. Furthermore, the emergency procedures (Emergency checklist: Both engines flame out) were not executed correctly by the flight crew following the successive flame-outs, resulting in a considerable and at this stage unwanted drag built up, contributing to a significantly worse aircraft drift down performance.[4]

---

4 In March 2009 an Italian court sentenced the Tuninter captain to ten years in jail for manslaughter. *Palermo, per la tragedia dell'Atr condannati pilota e tecnici di Tuninter*, La Republica, 23 March 2009.

As mentioned in the conclusion of the final investigation report, the accident was determined by a series of events (contributing factors), which were interlinked. The following factors were considered to be of major importance:

- Errors committed by ground mechanics when searching for and correctly identifying the fuel indicator.
- Errors committed by the flight crew: non-respect of various operational procedures.
- Inadequate checks by the competent office of the operator that flight crew were respecting operational procedures.
- Inaccuracy of the information entered in the aircraft management and spares information system and the absence of an effective control of the system itself.
- Inadequate training for aircraft management and spares information system use and absence of a responsible person appointed for managing the system itself.
- Maintenance and organization standards of the operator unsatisfactory for an adequate aircraft management.
- Lack of an adequate quality assurance system.
- Inadequate surveillance of the operator by the competent Tunisian authority.
- Installation characteristics of fuel quantity indicators (FQI) for ATR-42 and ATR-72 which made it possible to install an ATR-42 type FQI in an ATR-72, and *vice versa.*[5]

With respect to the technical investigation a dispute arose between the Tunisian and Italian authorities regarding competency. The Tunisian civil aviation authority, *in casu*, the *Directeur Général de l'Aviation Civile* (DGAC), refused the right of investigation by the Italian ANSV. In his letter of 6 December 2007 to ANSV the DGAC reiterated that the accident occurred in international waters and that thus the investigation should have been conducted by the State of Registry of the aircraft or the State of the Operator, in this specific case the Tunisian Republic. The DGAC further complained:

> Even though Italy, in accordance with Art. 38 of the Chicago Convention, had notified ICAO of non existence of any difference with Annex 13, during the investigation process, the ANSV has not observed International Civil Aviation Convention Annex 13 rules especially provision of Articles 5.3 and 5.25. (…) On the other hand, neither Annex 13 Paragraph 4.2 g and k (physical characteristics of the accident or serious incident area, as well as an indication of access difficulties or special requirements to reach the site) nor Annex 13 Paragraph 4.4 were observed by ANSV:
> 4.2 The notification shall be in plain language and contain as much of the following information as is readily available, but its dispatch shall not be delayed due to the lack of complete information:

5 *ANSV Final Report. Accident involving ATR 72 aircraft, registration Marks TS-LBB, ditching off the coast of Capo Gallo (Palermo-Sicily) 6 August 2005.*

(…)
g) position of the aircraft with reference to some easily defined geographical point and latitude and longitude:
j) an indication to what extent the investigation will be conducted or is proposed to be delegated to the State of Occurrence:
(…)
4.4 As soon as it is possible to do so, the State of Occurrence shall dispatch the details omitted from the notification as well as other known relevant information.
In this context it must be emphasized that some information is different from previously communicated by ANSV and Italian justice ones particularly when dealing with the impact and wreckage coordinates. One shall bear in mind the Annex 13 Paragraph 5.3 and 5.3.1:
5.3 When the location of the accident or serious incident cannot definitely be established as being in the territory of any State, the State of Registry shall institute and conduct any necessary investigation of the accident or serious incident. However, it may delegate the whole or any part of the investigation to another State by mutual arrangement and consent.
5.3.1 States nearest the scene of an accident in international waters shall provide such assistance as they are able and shall, likewise, respond to requests by the State of Registry.
Moreover and according to ANSV, it was impossible to stick to Annex 13 rules as judicial authorities did not recognize the right granted by Annex 13 Paragraph 5.25 and 5.27.
5.25 Participation in the investigation shall confer entitlement to participate in all aspects of the investigation, under the control of the investigator-in-charge (…)
5.27 A State which has a special interest in an accident by virtue of fatalities or serious injuries to its citizens shall, upon making a request to do so, be permitted by the State conducting the investigation to appoint an expert who shall be entitled to:
a) visit the scene of the accident;
b) have access to the relevant information;
c) participate in the identification of the victims;
d) assist in questioning surviving passengers who are citizens of the expert's State; and
e) receive a copy of the Final Report.

According to the above mentioned facts, it became obvious that ANSV, in total violation of the Chicago Convention, did not take into consideration

> the right of Tunisia as the State of Registry to conduct this investigation. Furthermore ANSV did not have any procurement or habilitation from Tunisian authority to lead this investigation.[6]

The following are specific remarks of the ANSV on the comments of the Tunisian DGAC:

> ANSV is the competent authority to conduct the investigation on ATR-72, TS-LBB, ANSV conducted it since the day of the accident. The supposedly rights of Tunisia as State of Registry and Operator to conduct the investigation is not acceptable. On the basis of available evidences, given by Italian Navy and Italian Maritime Authorities, the aircraft ditched within Italian territorial waters.[7]

On 14 August 2005 a Helios Airways Boeing 737-300 (Flight HCY 522 from Larnaca to Athens) with registration 5B-DBY crashed into a mountain near Grammatikos, 33 km northwest of Athens international airport (known as *Elefthérios Venizélos*), Greece. All 121 occupants on board were killed.

Before take-off the flight crew failed to switch the pressurization system mode selector to automatic and misinterpreted some crucial master caution and warning signals indicating abnormal pressurization system status, resulting in incapacitation of almost all occupants by hypoxia while flying at (high) cruising level. Owing to fuel exhaustion, the engines flamed out and the plane dived and eventually hit the ground uncontrolled.[8]

The Greek Air Accident Investigation and Aviation Safety Board (AAIASB), residing under the Hellenic Republic Ministry of Transport and Communications, investigated the accident in accordance with ICAO Annex 13, the Hellenic Public Law 2912/2001 and EU Directive 94/56. In its Final Report the Board indicated that it examined the evidence to determine the cause(s) of the accident and the analysis included examination of safety issues and safety deficiencies related to:

- Maintenance procedures and actions by the mechanics;
- Pilot training, procedures, and actions by the pilots;
- Organizational factors related to the Helios Airways flight operations and safety management systems;
- Organizational factors related to safety oversight of maintenance and flight operations by the Cyprus DCA, EASA, ICAO and JAA;

6 Letter from DGAC Tunisia Hamadi Ben Khaleb to M. Bruno Franchi, President ANSV, 6 December 2007.

7 ANSV Final Report, at 329.

8 According to CVR inputs and observations of the lead F-16 pilot, flying alongside the B-737, the male cabin attendant who managed to get in the captain's seat, tried to control the crippled plane, though without success.

> - Organizational factors related to the airframe manufacturer's presentation of maintenance and flight crew procedures and management of precursor incident information to amend such procedures in a timely manner, and;
> - Organizational factors related to the aviation industry's management of precursor incident information to develop accident prevention measures in a timely manner.
>
> As the result of the investigation, the Board identified numerous safety deficiencies which were communicated to the parties involved, some of which have already been resolved by safety actions taken by the responsible organizations. Additional safety deficiencies are addressed in the Board's report as safety recommendations for further actions to prevent future accidents.[9]

In the Final Report it became clear to the Board that Cyprus' international obligations were not being met:

> What became apparent from the Board's consideration of the situation at the Cyprus Department of Civil Aviation (DCA), and what was evident from the review of the audits/evaluations of DCA, was that the DCA, and the Safety Regulation Unit (SRU) in particular, lacked the required expertise to move forward, become independent, and fulfil the international obligations of Cyprus as contained in the Chicago Convention and its Annexes.
> A review of the audits and follow up audits of Cyprus DCA performed by ICAO, EASA and JAA, disclosed several important findings, which should have been remedied in the shortest possible time. No records were obtained that would have documented any remedial action considered, initiated or completed. It was of concern to the Board that there was no evidence of actions and enforcement by the international regulatory agencies to require timely implementation of an acceptable action plan, although they had clearly established that Cyprus' international obligations were not being met.[10]

Apart from the fact that the flight crew did not recognize and correct the incorrect system status and subsequently misinterpreted the reason for the warning horn and caution lights as well as the fact that the manufacturer's procedures should have contained enough redundancy to ensure the pressurization system was properly configured for the flight phase, organizational safety deficiencies existed within the operator's management

9 AAIASB Accident Investigation Report 11/2006 Helios Airways flight HCY 552, at 112.

structure and safety culture, as evidenced by diachronic findings in the audit prior to the accident, including:

a. Inadequate quality system;
b. Inadequate operational management control;
c. Inadequate quality and operations manual;
d. Cases of non-attendance of management personnel at quarterly management quality review meetings, as required;
e. Organization, management, and associated operational supervision not properly matched to the scale and scope of operations;
f. Inadequate monitoring of pilot certificates and training;
g. Insufficient involvement of management pilots in managerial duties, due to lack of time;
h. Incompletely updated training and duty records;
i. Lack of updating of some manuals and in part not fully in compliance with regulations;
j. Key management personnel at time performing the work of two positions;
k. Periods of vacant key management positions;
l. Inadequate remedial actions on audit findings, including level one findings, which could cause suspension of the AOC.

In addition, within the Cyprus DCA, organizational safety-related deficiencies existed at least since 1999 and continued to the time of the accident, despite the assistance and advisory service of the UK CAA. These deficiencies, although some of them were corrected, prevented the DCA from carrying out its safety oversight obligations within Cyprus.

In the Final Report the Board came to the following conclusion:

> Direct causes
>
> 1. Non-recognition that the cabin pressurization mode selector was in the MAN (manual) position during the performance of the:
>   a) Pre-flight procedure;
>   b) Before start checklist; and
>   c) After take-off checklist.
> 2. Non-identification of the warnings and the reasons for the activation of the warnings (cabin altitude warning horn, passenger oxygen masks deployment indication, Master Caution), and continuation of the climb.
> 3. Incapacitation of the flight crew due to hypoxia, resulting in continuation of the flight via the flight management computer and the autopilot, depletion of the fuel and engine flameout, and impact of the aircraft with the ground.

10 AAIASB Accident Investigation Report 11/2006 Helios Airways flight HCY 522, at 146.

Latent causes

1. The operator's deficiencies in organization, quality management and safety culture, documented diachronically as findings in numerous audits.
2. The Regulatory Authority's diachronic inadequate execution of its oversight responsibilities to ensure the safety of operations of the airlines under its supervision and its inadequate responses to findings of deficiencies documented in numerous audits.
3. Inadequate application of Crew Resource Management (CRM) principles by the flight crew.
4. Ineffectiveness and inadequacy of measures taken by the manufacturer in response to previous pressurization incidents in the particular type of aircraft, both with regard to modifications to aircraft systems as well as to guidance to the crews.

Contributing factors to the accident

1. Omission of returning the pressurization mode selector to AUTO after unscheduled maintenance on the aircraft.
2. Lack of specific procedures (on an international basis) for cabin crew procedures to address the situation of loss of pressurization, passenger oxygen masks deployment, and continuation of the aircraft climb.
3. Ineffectiveness of international aviation authorities to enforce implementation of corrective action plans after relevant audits.[11]

In general, the latent or underlying causes of accidents and serious incidents are imputing to or might give clues about structural safety defects or shortcomings in different systems such as the operator's safety management system, aircraft procedures training system or the civil aviation authority oversight system.

A profound technical investigation into the cause of an aviation accident or serious incident is primarily focusing on detecting and identifying these safety defects in order to be able to formulate and issue safety recommendations to prevent similar occurrences in the future.

Only two days later, on 16 August 2005, West Caribbean Airways Flight WCW 708 (aircraft type: McDonnell Douglas MD-82, registration: HK-4374X), on a charter flight from Panama City Tocumen international airport to Fort de France-Lamentin international airport on the French Caribbean island Martinique, crashed near Machiques in a mountainous region in the northwestern part of Venezuela, killing all 152 passengers and 8 crew members, and made it the deadliest civil aviation accident in 2005. This particular accident, as nearly all the passengers, were French citizens of Martinique, and the preceding accidents in Europe had a tremendous impact on European society.

The *Comité de Investigación de Accidentes Aéreos* (CIAA) of Venezuela led the investigation on the causes of the accident, while the U.S. NTSB and the French *Bureau d'Enquêtes et d'Analyses pour la Sécurité de l'Avion Civile* (BEA) also took part.

Factual information released by CIAA after initial examinations of the wreckage on sight revealed:

- Ground scars indicate that the airplane impacted in a nose up and slight right roll attitude.
- Wreckage was distributed over a triangle shaped area that was approximately 205 meters long and 110 meters at its widest point.
- Both engines exhibited indications of high-speed compressor rotation at the time of ground impact.
- The engine inlets, empennage and wing leading edges showed no sign of pre-impact damage.
- The horizontal stabilizer was found at about the full airplane nose up position (about twelve units nose up).

The following events are recorded on the Flight Data Recorder:

- About 90 seconds after reaching Mach 0.76 at Flight Level 330 the airspeed began to steadily decrease. About 3 minutes and 30 seconds from the end of recording, the Mach number reached about 0.60. The autopilot was then disengaged and the airplane started to descendfrom FL 330.
- As the airplane descended past about FL 315, the airspeed continued to decrease and the right engine – Engine Pressure Ratio – (EPR) decreased to about flight idle.
- The airplane descent rate increased after passing through FL 310.
- The airspeed reached a minimum of about 150 indicated air speed (IAS) knots at about FL 250. The altitude alert warning had activated, followed by the stick shaker and the aural stall warning alert.
- Right engine EPR stayed at approximately flight idle through the descent and even increased several times, including shortly before the end of the recording.
- Once the airplane started to descend, the horizontal stabilizer moved in increments to about twelve units nose up while descending through FL 200.

The Cockpit Voice Recorder provided some useful information, especially during the last eight minutes of the recording. The time interval from the first sound similar to a stick shaker (warning to imminent stall condition) to the end of the recording is approximately

11 AAIASB Accident Investigation Report 11/2006 Helios Airways flight HCY 552, at 159.

2 minutes and 46 seconds. The entire descent from FL 330 had taken approximately 3 minutes and 30 seconds. Approximately 8 seconds before the end of the recording, a ground proximity warning system (GPWS) warning started to be heard and continued to the end of the recording.[12]

The following are new findings concerning the MD-82:

- The plane made a moderately overweight take-off from Panama City.
- The flight crew apparently took the most direct route to make up delay time; however, they were about to encounter an area of thunderstorms with moderate turbulence and icing conditions.
- Data derived from the CVR revealed that the pilots were requesting ATC for clearance to deviate from the actual track because of weather, and that the pilots were arguing about the anticipation of a possible penetration of severe weather, including potential ice built up and the obvious use of engine nacelle and wing anti-ice, which as a known consequence, however, would cause reduced engine thrust and lift production.
- The intermediate climb from FL 310 to FL 330 was initiated to avoid icing conditions; however, owing to the use of the anti-ice system and the weight of the aircraft, this critical manoeuvre was performed outside the aircraft's flight envelop (maximum engine thrust was insufficient at that time to reach and maintain speed and higher altitude). At level off the auto-throttle system was slow to add the required power to meet the pre-selected speed. The flight crew did not realize this. The autopilot, however, started to raise the nose to compensate. Speed was decreasing significantly. The flight crew requested lower flight levels to keep the aircraft flying. After about 6 minutes, the autopilot, unable to handle the aircraft anymore under these circumstances, was disconnected and 2 minutes later the aircraft stalled (stall warning activation).
- The flight crew never recognized the low power setting and subsequently the irrevocable stall. During the deep-stall manoeuvre the flight crew attempted to counteract the stall condition but their recover technique input was completely opposite.

The following are some conclusions of the final report:

- Poor situational awareness of the flight crew.
- Inadequate performance of the flight crew regarding flying in adverse weather, especially monitoring of flight instrument indications and aircraft performance degradation.
- Not familiar with recent MD-82 safety bulletins.
- The deplorable financial state of West Caribbean Airways contributed to far from optimum flight operations, training and safety management.

12 NTSB Advisory factual information, 22 November 2005.

- Recommendation to improve CRM training and Simulator training with emphasis on memory items in emergency situations, situational awareness, instrument and system monitoring.[13]

These (and previous) fatal aircraft accidents, killing a large number of European travellers, have undoubtedly had an impact on the timely production of Regulation (EC) No 2111/2005. But in essence, this Regulation is part of a legislative process pursuing an efficient and coherent approach to reinforcing aviation safety in the Community.

8.3 EU BLACKLIST

With a view to reinforcing aviation safety and further ensuring the safety of air transport passengers and to some extent the safety of European citizens living around international airports, the European Commission established, in 2005, a Community blacklist (the EC list or EU blacklist) of air carriers subject to an operating ban within the European Union.[14] As a result, the respective aircraft operators and aircraft identified in this list are not allowed to operate within the European airspace.

Regulation (EC) No 2111/2005 entered into force on 16 January 2006, and Commission Regulation (EC) No 473/2006 and No 474/2006 establishing and laying down implementing rules for the Community list of air carriers that are subject to an operating ban within the Community referred to in Chapter II of Regulation (EC) 2111/2005, the first such list based upon legally established common criteria, which in turn are based on the relevant safety standards, were adopted on 22 March 2006, coming into force on 24 March 2006.

Since then, the list has been updated frequently.[15] The updates have been the result of continuous monitoring and assessment of the performance of air carriers, irrespective of their nationality and of the geographic coverage of their operations.

---

13 *Informe Final Accidente Aéro Aeronave Modelo MD-82 Matrícula HK-4374X Vuelo No WCW 708, Junta Investigadora de Accidentes de Aviación Civil (JIAAC) del Ministerio del Poder Popular para Transporte y Communicaciones de la Repùblica Bolivariana de Venezuela.*

14 Regulation (EC) No 2111/2005 of the European Parliament and of the Council of 14 December 2005 on the establishment of a Community list of air carriers subject to an operating ban within the Community and on informing air transport passengers of the identity of the operating carrier, and repealing Article 9 of Directive 2004/36/EC, *Official Journal of the European Union L* 344/15, 27 December 2005. *See under* Article 2(g) of this Regulation the definition of an operating ban: the refusal, suspension, revocation or restriction of an air carrier's operating authorization or technical permission for safety reasons, or any equivalent safety measures in respect of an air carrier which has no traffic rights in the Community but whose aircraft might otherwise be operated in the Community under a lease agreement.

15 From June 2006 till November 2011 (Commission Implementing Regulation (EU) No 1197/2011 of 21 November 2011 amending Regulation (EC) No 474/2006 establishing the Community list of air carriers which are subject to an operating ban within the Community).

Basically, EU member States have the right to publish a list of all air carriers banned from their airspace or subject to traffic right restrictions for safety reasons. These lists are made available to other member States and to the European Commission, who, in fact, will publish a consolidated list, which is universally accessible. At least every three months, the European Commission must assess whether the blacklist needs an update to include or exclude certain carriers.[16] To be able to establish an update, the member States and EASA's Air Safety Committee must notify the European Commission of all relevant data. In 2006 alone, the EU banned 92 air carriers, the vast majority of them from Africa.

The need to adopt Regulation (EC) No 2111/2005 resulted from the weaknesses in the enforcement of the internationally agreed safety standards laid down in the Chicago Convention and its Annexes under the auspices of ICAO, where non-European air carriers are concerned and the European Community air safety *acquis,* where Community air carriers are concerned.

For this reason the Regulation provides for the imposition of a total or a partial ban where substantiated evidence from objective, transparent and quantifiable criteria show that ICAO safety standards are not being followed by air carriers. This Regulation is designed to give air transport passengers the right to be informed of the identity of the air carrier operating the flight they have booked and the right of reimbursement or transfer to another air carrier in case of an imposed cancellation of a flight booked with an air carrier added to the EU blacklist.[17]

At the same time this Regulation will reinforce the obligation on EU member States to communicate aviation safety-related information to all the member States and the European Commission that may be essential to updating the blacklist.[18]

This common approach method is standard practice in the Community, but in exceptional cases, member States should be allowed to take unilateral measures. It should be possible for a single member State to immediately impose an operating ban in respect of its own territory, taking into account the common criteria, when confronted with an unforeseen or urgent safety problem threatening that member State.[19]

In fact, air carriers are placed on the Community list if strong evidence is gathered regarding unremitting or persistent poor performance concerning safety criteria. Findings of SAFA inspections, carried out at Community airports, will provide an indication of non-compliance with minimum international safety standards. The defects are clearly reflected in badly maintained and obsolete aircraft, mainly due to the inability or

16 Article 4.2 Regulation (EC) No 2111/2005.

17 Articles 10, 11 and 12 of Regulation (EC) No 2111/2005.

18 Article 4.3 of Regulation (EC) No 2111/2005.

19 Regulation (EC) No 2111/2005 under (8) and Article 6 Exceptional Measures.

negligence of air carriers in question to take corrective actions with respect to any identified deficiencies. Inability in this particular case is often a result of inadequate or incorrect regulatory oversight or poor cooperation by the responsible national civil aviation authorities, especially in third countries where the safety levels depend on the effectiveness of applicable oversight procedures. Furthermore, related safety information to update or complement the EU blacklist may include audit reports drawn up by ICAO and safety inspections within the framework of USOAP performed by the civil aviation authorities of the Chicago Convention contracting States as well as operating bans imposed by non-EU or third countries.

In addition, substantiated information concerning aircraft accident investigation or other serious incident-related information will in any form contribute to a decision to amend the blacklist by including or removing an air carrier.[20]

Only the European Commission or a member State may ask for the blacklist to be updated. If an air carrier considers that it should be removed from the blacklist because it complies with all relevant safety standards, it may submit a request to the European Commission or a member State, either immediately or through its civil aviation authority.[21]

The air carrier in question and/or its safety oversight authority have to present conclusive evidence to substantiate its request for being withdrawn from the blacklist by the European Commission after an assessment and advisory opinion from the Air Safety Committee. This procedure should allow for decisions to be taken swiftly, without undue delay, or even urgently where appropriate, in order to provide adequate, transparent and up-to-date safety information to air passengers and to guarantee that air carriers that have remedied identified safety deficiencies are removed from the blacklist as soon as practicable.[22]

As long as an air carrier is banned totally, it cannot operate with its aircraft and personnel within the European airspace. However, a banned carrier is still able to fly its aircraft into the Community for regular maintenance or for reasons of resolving safety deficiencies in European facilities, naturally without carrying any revenue passengers or other payload (cargo, baggage).

On the other hand, banned air carriers, permitted to exercise their acquired traffic rights, can use aircraft and personnel from other air carriers based on wet-lease contracts. Obviously, these letting carriers must fully comply with international safety requirements.[23]

---

20 *See* Annex to Regulation (EC) No 2111/2005 under (1).

21 Regulation (EC) No 2111/2005, Article 4 Updating of the Community list. *See also* Commission Regulation (EU) No 1071/2010 under (3).

22 Regulation (EC) No 2111/2005 under (6) and Article 8.2.

23 *See* Regulation (EC) No 2111/2005 under (5).

Regulation (EC) No 2111/2005 requires all air transport travellers to be informed of the identity of the air carrier that will actually operate any flight to or from an EU international airport, as part of a journey, subject to a contract of carriage, which starts or ends in the European Union. The identity of the air carrier or carriers actually operating the flight is an essential piece of information. If the operator actually changes, the passenger concerned should be informed of the identity of the new air carrier at the earliest opportunity:

> Wherever the operating air carrier or carriers is or are changed after reservation, the air carrier contractor shall, irrespective of the reason for the change, take immediately all appropriate steps to ensure that the passenger is informed of the change as soon as possible. In all cases, passengers shall be informed at check-in, or on boarding where no check-in is required for a connecting flight.[24]

The rules establishing the Community list of banned air carriers apply to all carriers irrespective of their nationality. In a report to the European Parliament and the Council published on 11 January 2010, the European Commission concluded that its blacklist, which was initially considered controversial, is now regarded internationally as an effective tool in ensuring a high level of safety to the benefit of the travelling public by the enforcement of the relevant international aviation safety standards.

The criteria for consideration of an operating ban for safety reasons are laid down in an Annex to Regulation (EC) No 2111/2005. Whether an air carrier or all the carriers certified in the same State are eligible for action at the Community level to be partially or totally banned shall be assessed according to the following facts and observations in the context of whether or not to meet relevant safety standards:

1. Verified evidence of serious safety deficiencies on the part of an air carrier:
   a. Reports showing serious safety deficiencies, or persistent failure by the carrier to address deficiencies identified by ramp inspections performed under the SAFA programme previously communicated to the carrier.
   b. Serious safety deficiencies identified within the framework of the provisions for the gathering of information in Article 3 of Directive 2004/36/EC on the safety of third-country aircraft.
   c. Operating ban imposed on a carrier by a third country because of substantiated deficiencies related to international safety standards.

24 Article 11.3 (Information on the identity of the operating air carrier) Regulation (EC) No 2111/2005. *See also* Regulation (EC) No 2111/2005 Article 2 Definitions, under (e) operating air carrier: an air carrier that performs or intends to perform a flight under a contract of carriage with a passenger, or on behalf of another person, legal or natural, having a contract of carriage with that passenger.

   d. Substantiated accident-related information or serious incident-related information indicating latent systemic safety deficiencies.
2. Lack of ability and/or willingness of an air carrier to address safety deficiencies as demonstrated by:
   a. Lack of transparency or adequate and timely communication on the part of the carrier in response to an inquiry by the civil aviation authority of a member State regarding the safety aspect of its operation.
   b. Inappropriate or insufficient corrective action plan presented in response to an identified serious safety deficiency.
3. Lack of ability and/or willingness of the authorities responsible for the oversight of an air carrier to address safety deficiencies as demonstrated by:
   a. Lack of cooperation with the civil aviation authority of a member State by the competent authorities of another State, when concerns about the safety of the operation of a carrier licensed or certified in that State have been raised.
   b. Insufficient ability of the competent authorities with regulatory oversight of the carrier to implement and enforce the relevant safety standards. Particular account should be taken of the following:
      - audits and related corrective action plans established under ICAO's Universal Safety Oversight Audit Programme or under any applicable Community law;
      - whether the operating authorization or technical permission of any carrier under the oversight of that State has previously been refused or revoked by another State;
      - the Air Operator's Certificate has not been issued by the competent authority of the State where the carrier has its principle place of business.[25]

Any one of these common criteria, set up for imposing a total or partial operating ban, on its own, or in combination with others, may be used as the basis for proposing the inclusion or exclusion of an air carrier, or a number of air carriers, in the EU blacklist. The criteria as shown in Annex A are grouped in three areas:
a. Objective evidence showing deficiencies on the part of the air carrier;
b. Lack of ability or willingness by an air carrier to address safety deficiencies; and
c. Lack of ability or willingness of the civil aviation authority with responsibility of oversight over the air carrier(s) in question to address safety deficiencies.[26]

25 Annex to Regulation (EC) No 2111/2005.

26 *See also*: Commission Staff Working Document SEC (2009) 1735 final, (COM (2009) 710 final), on the report from the Commission to the Council and the European Parliament regarding the application of Regulation (EC) No 2111/2005, Part 2. Analysis of the common criteria for deciding the imposition of an operating ban.

These common safety criteria for consideration of an operating ban focus mainly on the safety-related information available in the member States and in the European Commission and EASA.[27] During the adoption process of Regulation (EC) No 2111/2005, these transparent and objective criteria were drawn up on the basis of work undertaken by the Air Safety Committee of experts, reflecting the long-standing and vast air safety experience of EU member States, next to the expertise of international aviation-related organizations in this particular field.

An air carrier, using poorly maintained, antiquated or obsolete (unsafe) aircraft, or which is dysfunctional to rectify repetitive shortcomings identified during SAFA ramp inspections at Community airports, will initially be subject to an analyzing process by the European Commission on a case-by-case basis. During this process, the civil aviation authority of the air carrier concerned, responsible for overseeing its poorly performing air carrier but for various reasons remaining in default to ensure that international safety requirements at all times are respected, will be addressed. The European Commission will take a radical and far-reaching decision to impose a complete or partial ban on a poorly performing air carrier only after it has analyzed all the various factors and after having given the air carrier in question the right to explain its position, to submit any relevant documents, the right to a fair hearing, and after having held detailed consultations with the safety oversight authority of the air carrier concerned and with the Air Safety Committee of experts.

With regard to a ban imposed by a single member State, there is scope for continuing to act at the national level in exceptional cases such as emergencies or in response to a safety issue specifically affecting that member State. When an air carrier is considered unsafe or risky and banned by a single member State, there is, however, an obligation, based on a general principle, to examine this serious measure at the Community level with a view to applying it throughout the European Union.

As provided under Regulation (EC) No 2111/2005, the blacklist comprises two annexes: Annex A containing air carriers whose commercial operations are subject to a total ban in the territory of EU member States, and Annex B containing air carriers whose commercial operations are subject to a partial ban or operating restrictions within the territory of EU member States.[28]

With respect to the notification of the intention to propose inclusion of non-complying air carriers and their respective regulatory authority in the EU blacklist and the facts

---

27 *See also* Annex A to Commission Regulation (EC) No 473/2006 of 22 March 2006 laying down implementing rules for the Community list of air carriers which are subject to an operating ban within the Community referred to in Chapter II of Regulation (EC) No 2111/2005 of the European Parliament and of the Council, *Official Journal of the European Union L* 84/8, 23 March 2006.

28 Report from the Commission to the Council and the European Parliament COM (2009) 710 final of 11 January 2010, at 2.

leading to such a decision, Article 7 (Rights of defence) of Regulation (EC) No 2111/2005 and Article 4 (Exercise of carriers' right to defence) of Regulation (EC) No 473/2006 provide for a profound and democratic (fair and transparent) procedure.

In case of an imminent decision under these articles, the European Commission shall disclose to the respective air carrier the essential facts and considerations that form the basis for such a drastic decision. The air carrier concerned and its respective regulatory authority shall be given an opportunity to submit any further documents and comments that, in their opinion, will be pertinent or appropriate in their defence. Moreover, in practice a carrier will be given the opportunity of a hearing before the Air Safety Committee prior to a final decision being taken on its inclusion in the EU blacklist, even in case of urgency.[29] As of 16 February 2006 the European Commission established for the first time a Community blacklist based on common criteria and information received from the EU member States. The EU member States are obliged to communicate to the European Commission the identity of air carriers that are subject to an operating ban in their respective territories at that particular time, together with the reasons for the imposition of such bans and other relevant information.[30]

Regulation (EC) No 2111/2005 stipulates, under Articles 10 and 11, that all passengers must be informed about the identity of the air carrier (or carriers) operating a flight for which they are making a reservation when such flight is departing from or arriving at a Community airport or is part of an itinerary that begins or ends in the territory of a EU member State. Moreover, wherever the operating air carrier or carriers is or are changed after reservation, passengers must be informed as soon as possible regarding the identity of the new air carrier (or carriers) operating any part of the itinerary in question. This might be extremely important for EU citizens travelling outside the European Union with an air carrier (or carriers) subject to a total or partial operating ban and subsequent inclusion in the EU blacklist. These risky carriers are banned from operating within European airspace, but can still operate with impunity in other parts of the world, although the scope of the EU blacklist is now internationally regarded as an effective tool in ensuring a high level of safety performance by air carriers through the enforcement of applicable safety standards.

8.4 Monitoring

The European ban is a temporary measure with a proportional level. If a banned air carrier can demonstrate that it has addressed all preliminary identified serious safety deficiencies in a sustainable way, and subsequently identifiable, by means of monitoring its

29 *See* also Article 4.3 of Regulation (EC) No 473/2006.

30 *See* Regulation (EC) No 2111/2005 Articles 3.3 and 3.4, Establishment of the Community List.

performance, shall operate in conformity with the relevant safety standards, it will be forthwith removed from the blacklist. Monitoring performance is achieved by means of consistent reporting on the part of the air carrier and its respective oversight authority. The Air Safety Committee not only has an important role in advising with regard to inclusion, but also when it comes to ensure a removal of a banned air carrier from the blacklist. Apart from its own research, the Committee has continuous access to information from the authorities of member States and non-EU States, as well as from dedicated international organizations, to be able to comprehensively audit and confirm the completion of remedial and corrective measures taken by the banned air carrier as identified in rectifying action plans.

By means of monitoring and processing relevant information, a clear and substantiated opinion regarding the level of safety of the air carrier concerned is accomplished by the Committee in order to advise the European Commission.

Being removed from the Community list means that the safety deficiency or deficiencies that gave rise to the inclusion of the air carrier concerned have been remedied and that no other reasons, on the basis of the common criteria, to maintain the air carrier on the blacklist, are present.[31]

On the other hand, removal of air carriers from the Community list could be caused by their full or partial cessation of operations and the rigorous revocation of their AOCs by their regulatory authorities. In many cases cessation of operations should be regarded as a direct autonomous result of the EU ban or subsequent bans.

The regularly updated and nowadays well-respected Community list serves as a paradigm for a number of States beyond the European Union. They voluntarily follow the comprehensive blacklist by considering the included air carriers as not complying with the relevant safety standards as these States have adopted, and therefore regarded by these States as banned from operating into their territory.

A rather new trend concerning implementation of corrective actions by suspected air carriers has developed lately. Instead of the imposition of a ban by the European Commission on air carriers not complying with safety standards, a proactive process, whereby cases are solved through a cooperative exchange between the European Commission and aviation authorities concerned, assists in lifting severe restrictions on operations of air carriers concerned. The procedure is as follows: States, when presented with verified evidence by the European Commission that serious safety deficiencies on one or more of their air carriers prevail, could immediately suspend the AOCs of the air carriers concerned or directly impose restrictions on their operations into the airspace of EU member States.

31 Article 4 (Updating of the Community list), 1. (b) of Regulation (EC) No 2111/2005.

By stringent supervising and monitoring by the authorities regarding remedial and adequate corrective actions on the part of the air carriers, in order to prevent recurrence of previously observed safety deficiencies, these State authorities are permitted, after consultation with the European Commission, to lift any suspension or severe restrictions, and thus to avoid an extremely unpopular punitive measure such as a ban. In fact this method will tend to be a more pre-emptive procedure.

Information derived from ICAO plays a major role in the EU's blacklist policy. States that expressed having significant difficulties in fulfilling their safety oversight commitments during the ICAO USOAP audit process, especially in failing to implement ICAO SARPs, resulting in persistent safety deficiencies in such a way as to aggravate ICAO's concern, particularly have observed their air carriers being subject to a total operating ban in the Community. To transform this serious failure for the better, the European Commission cooperates closely with non-EU countries as well as international aviation organizations such as ICAO (ICAO COSCAP projects) to verify correct implementation and compliance and therefore enhance international aviation safety oversight in general.[32] COSCAP programmes may supplement safety oversight capabilities of States in areas where these States have specific needs. The programmes may also serve as an instrument for States to make a transition to a full-grown Regional Safety Oversight Organization, where the circumstances support the establishment of such a self-sustaining (sub)-regional organization.

The ICAO COSCAP programme is focusing mainly on flight operations and airworthiness safety oversight, but as the scope of the USOAP audit has expanded, COSCAP has accordingly widened its field of activity to provide cooperation in the area of aerodromes, accident and incident investigation, air traffic management and dangerous goods by air. The ICAO COSCAP has proven to be an instrument to enhance aviation safety in the broadest sense, especially for the benefit of regional State safety oversight and, in consequence, to prevent the imposition of potential EU total or partial regional air carrier operating bans. On the other hand, the ability to comply with internationally agreed safety standards by regional air carriers in the convalescence phase following the removal from the EU blacklist, clearly demonstrates the successful application of results-oriented,

32 COSCAP: Cooperative Development of Operational Safety and Continuing Airworthiness Programmes. There are various COSCAP programmes in different regions: ACSA (Central American Agency of Aeronautical Safety), BAGASOO (Africa region), CASSOA (Civil Aviation Safety and Security Oversight Agency – Africa region), CASSOS (Caribbean Safety and Security Oversight System), CEMAC (Africa region), Gulf States (Middle East region), North Asia (Asia Pacific region), SADC (Africa region), SEA South East Asia (Asia Pacific region), South Asia (Asia Pacific region), UEMOA (Africa region), EASA (Europe region), IAC Interstate Aviation Committee (Eastern Europe region), PASO Pacific Aviation Safety Office (Asia Pacific region) and SRVSOP Regional Safety Oversight Cooperation System (North America, Central America, South America and Caribbean region).

remedial efforts by these earlier banned carriers and regional State safety oversight authorities, evaluated on the basis of substantive and objectively presented criteria.

The regional COSCAP programmes are providing for dedicated forums for promoting continuing dialogue, coordination and cooperation in matters relating to flight safety among the well-developed and smaller participating civil aviation administrations and creating an environment for harmonization, standardization and advancement in safety oversight policies, procedures and regulations. The regional COSCAP participant States are supported mostly by EU institutions, U.S. Governmental bodies such as the FAA, the Canadian Government, IFFAS, Boeing and Airbus, all regarded as partners. The COSCAP programmes provide, in addition, efficient and workable methods to support the inspection of operators, aircraft and training facilities and education and training for a large number of safety oversight personnel. Another important objective in different regions involves a proactive approach regarding effective prevention of aviation accidents.

8.5 EU Policy

Two prominent policy aspects are brought up by the European Commission with respect to irregularities or shortcomings in the blacklist methodical system. More detailed legislative provisions must clarify the way member States should act when they are affected by an attempted breach of the EU ban. In addition, the European Union should closely monitor and record decisions taken by certain countries to restrict, for safety reasons, the AOCs of their carriers that are operating flights into the European Union. Strengthening of the SAFA programme by further refining the existing relevant legal instruments will enable the member States to detect safety deficiencies in a more sophisticated and reliable way and to react more promptly to mitigate any risk, even by a minimum number of inspections.

Information derived from aircraft accident and incident investigations is increasingly important as a tool to update the Community list. Identification of causal factors of an aircraft accident or serious incident is best accomplished through a properly conducted comprehensive investigation.

The European Commission closely monitors the occurrence, technical investigation and reporting of accidents and serious incidents. Where substantiated information is available, in particular final reports of accident investigations as well as any other relevant safety details, any negative safety trends or more immediate safety hazards requiring urgent action could be identified.

Advanced aviation technology has led to aircraft systems to be more complex, which undoubtedly implies that accident investigation needs substantially more specialization, cooperation and diversified expertise and resources. As the civil accident investigation and occurrence reporting system in the enlarging European Union was concentrated only in a few experienced member States, and its capacity generally was functioning below optimum efficiency, this institutional and legal framework regarding accident investigations, given the increasing responsibility, has been given a reinforcing structure following a proposal of the European Commission in 2009 to create a network of accident investigation bodies.[33]

One of the policy options available to the European Commission to address the inefficiencies associated with the mutual assistance between the member States in the field of air accident investigation was the establishment of a comprehensive European Network of Civil Aviation Safety Investigation Authorities.[34]

As a result, this important step in the right direction shall move aviation accident investigation from the State to the European level. The objective of the Network will be to develop common approach procedures to aviation accident investigation in the European Union, to strengthen the investigation capacity and cooperation and independence of national competent investigation bodies (aviation safety boards) and recommend EU institutions on aviation accident investigation and prevention matters.

The establishment of the Network, consisting of a group of already existing permanent air accident investigation authorities in the EU member States, is envisaged by Regulation (EU) No 996/2010 of the European Parliament and of the Council of 20 October 2010 on the investigation and prevention of accidents and incidents in civil aviation and repealing Directive 94/56/EC.

One of the coordinating tasks of the Network is to preserve the highest safety standards and to inform the public by ensuring that relevant information useful for the improvement of aviation safety is shared, reported, collected, stored, protected and disseminated. Accident-related information regarding the resolution of safety concerns is a major source, next to various audit reports and corrective actions following SAFA ramp checks, with respect to the evaluation framework regarding updating of the EU blacklist.

The EU blacklist undoubtedly is an effective and successful tool, although it has some restrictions with respect to inclusion. Inclusion on the EU blacklist depends on available and verifiable information, and it constitutes an operating ban only to the European region. However, banned air carriers are usually able to continue to fly without any restraint to other regions in the world. Therefore the European Commission strongly supports the promotion of exchanging relevant verifiable and reliable information at the international level to establish and maintain the highest possible level of aviation safety worldwide.

This can be accomplished only by a variety of measures to be able to effectively enforce compliance with ICAO safety standards by contracting States and their air carriers.

33 COM (2009) 611 final, Proposal for a Regulation of the European Parliament and of the Council on investigation and prevention of accidents and incidents in civil aviation.

34 COM (2009) 611 final, Articles 7 and 8.

Appropriate measures are technical assistance and bilateral partnership agreements with civil aviation authorities to manage oversight and organizational shortcomings, especially if significant safety concerns are present and audit reports indicate a considerable lack of implementation of international safety standards.

> The Commission intends to strengthen its ongoing cooperation with strategic partners in order to facilitate the exchange of safety data and establish a network of trusted interlocutors for evaluating not only the safety aspects of various organizations involved in the manufacture, operation, maintenance and training, but also the application of safety programmes at State level. Such cooperation should also include the exchange of comparable safety data relating to ramp inspections of aircraft, the objective being to align as much as possible the overall format of the reporting system of safety data to improve the capability of data used.[35]

However, if cooperation in any way is not sufficient to mitigate the risks to safety, banning air carriers will remain the necessary course of action.

By closely monitoring the performance of suspected air carriers and evaluating the results of ICAO USOAP audit reports evidently revealing numerous and serious deficiencies, the European Commission intends to continue to ensure that air carriers are limited or even not permitted to fly within Community airspace until their authorities or operators can guarantee total conformity with ICAO standards.

Through close cooperation with ICAO, which should intensify the coordination of applied assistance activities (implementation of State Safety Programmes) and propagate sharing of safety information and open publication of serious safety concerns, the European Commission intends to create an atmosphere where all contracting States shall respect the international safety standards, thus ensuring a high level of aviation safety throughout the world and not only where legal tools, such as Regulation (EC) No 2111/2005, apply.[36] This cooperation eventually must lead to a worldwide banning list. Several other States outside the European Union have similar mechanisms for imposing operating bans on air carriers because of non-compliance with safety standards. Through bilateral specific safety agreements the European Union and these non-EU States have established close cooperation with respect to the exchange of important safety information and access to

---

35 COM (2009) 710 final, Report from the Commission to the Council and the European Parliament on the application of Regulation (EC) No 2111/2005, at 9.

36 The State Safety Programme (SSP) is a system for the management of safety within a State. The SSP provides an enabling platform for States and civil aviation organizations to apply two basic safety management principles to the discharge of their safety responsibilities: Safety Risk Management (SRM) and Safety Assurance (SA). *Source*: HLSC, Montreal 29 March-1 April 2010 (ICAO Doc 9935).

their respective safety oversight and enforcement systems. There is also close cooperation with those civil aviation authorities with which the European Union is negotiating safety-specific agreements or comprehensive aviation-related agreements, to ensure that safety information regarding the performance of air carriers operating to the European Union and these States will be exchanged in a timely manner in order to allow for informed decisions on the continuation of their operations.

If an operating ban is imposed by a non-EU State, the European Union will monitor the situation closely and analyze this operating ban, in close harmony with the relevant regulatory authorities, to be able to assess whether circumstances dictate to impose a similar ban in Europe. In the opposite case, the lifting of a ban by a regulatory authority outside the European Union, but also in the event of an EU-initiated lifting, active coordination with the relevant regulatory authorities will be at issue.[37]

So far the EU blacklist has functioned as a strong incentive to air carriers and civil aviation authorities to continuously improve aviation safety. Through regular exchange of information with ICAO and other related organizations, the European Commission has been able to further adjust and refine its judicious understanding of the safety situation in various regions in the world and to adapt its technical assistance projects as necessary. The European Commission remains committed and engaged in several technical assistance projects, notably in the African region, to help civil aviation authorities, with the support of EASA, to overcome their organizational, administrative and operational issues in the most effective way. In the course of the application of Regulation (EC) No 2111/2005 it has become evident how many EU member States, EEA partners and Switzerland, predicated their safety decisions upon the results of ICAO safety audits carried out in the detailed framework of the USOAP programme. In fact it is a clear indication that the international community should consistently follow ICAO SARPs.

The European Commission strongly believes that encouraging respect for these safety standards on a global scale will be a task for ICAO, within the remit of its competencies under Article 54 of the Chicago Convention.[38]

In 2009 the European Transport Commissioner Antonio Tajani proposed to expand EU safety standards beyond European borders to improve aviation safety on a wider scale. In essence the proposal included a global instrument like an air carrier blacklist similar to the European Union's scheme. This new approach was initiated in the wake of the Comoros Islands accident involving a Yemenia Airlines Airbus A310-300 flight 626 from

37 *See also* Commission Staff Working Document. On the report of the Commission to the Council and the European Parliament regarding the application of Regulation (EC) No 2111/2005 regarding the establishment of a Community list of air carriers subject to an operating ban within the Community and informing air transport passengers of the identity of the operating air carrier, and repealing Article 9 of Directive 2004/36/EC, at 4.

38 COM (2009) 710 final, 3. Worldwide banning of unsafe air carriers, at 10.

Sana'a, Yemen, to Moroni, Comoros, on 30 June 2009, while on final approach to Prince Said Ibrahim international airport, killing 152 passengers and crew members.[39] The air carrier in question was already being closely monitored by EU authorities.

ICAO Council President Roberto Kobeh Gonzalez responded that using the blacklist concept was not a global level solution, but instead ICAO prefers a more preventive approach to enhance aviation safety.[40] "Discouraging passengers from going on particular aircraft or particular air carriers on certain routes will not necessarily reduce accidents. At present, I cannot be in favour of a world blacklist because I don't have a mandate from the States of ICAO to do this, but I fully agree with Commissioner Tajani we have to work together to improve safety".[41]

> The aims and objectives of the Organization are to develop the principles and techniques of international air navigation and to foster the planning and development of international air transport so as to:
> a. Insure the safe and orderly growth of international civil aviation throughout the world;
> b. Encourage the arts of aircraft design and operation for peaceful purposes;
> c. Encourage the development of airways, airports, and air navigation facilities for international civil aviation;
> d. Meet the needs of the peoples of the world for safe, regular, efficient and economical air transport;
> e. Prevent economic waste caused by unreasonable competition;
> f. Insure that the rights of contracting States are fully respected and that every contracting State has a fair opportunity to operate international airlines;
> g. Avoid discrimination between contracting States;
> h. Promote safety of flight in international air navigation;
> i. Promote generally the development of all aspects of international civil aeronautics.[42]

The key objective of ICAO, as a global co-ordinating civil aviation organization, is to help contracting States to reduce their accident rates by assisting in safety oversight and to promote compliance with international safety standards. "Our task is now to develop and

39 First loss of life in a Yemenia accident in 36 years, according to the Aircraft Crashes Record Office, Geneva-Switzerland.

40 <www.flightglobal.com>, 'ICAO Confirms Opposition Against a Global Airline Blacklist', 9 July 2009.

41 *ICAO Rejects Global Blacklist*, Air Transport World, 13 July 2009.

42 Article 44 of the Convention on International Civil Aviation (ICAO Doc 7300).

implement more flexibility and targeted strategies to help States with specific challenges in reducing their accident rate and thereby contribute to an even safer global air transport system, while at the same time reinforcing safety in others," according to Mr. González, President of the Council of ICAO.

Meanwhile, the European Commission, assisted by EASA, is working in close cooperation with non-EU countries, lacking the capability to catch up with safety management, regulatory and oversight, in order to bring them at a higher level of aviation safety by prevention projects and adequate technical assistance within the framework of Regulation (EC) No 2111/2005. This initiative is intended to support positive *ex ante* prevention actions as well as *ex post facto* technical cooperation to assist efforts made by States at implementing corrective action plans to address numerous identified serious safety deficiencies.

In addition, the European Commission will launch, under the policy proposal of a dual strategy pattern, parallel to the above-mentioned activities, the formulation of more detailed provisions through an amendment to the implementing legislation and improvements in the functioning of the EC SAFA programme by further refining the already existing legal instruments to mitigate any risks. A key aspect shall be the accurate analyzes and prompt exchange of information stored in the EASA database.

> The new way of doing business will allow proactive identification and response to emerging safety issues before they result in accidents or incidents. This involves understanding the risks associated with the many facets of today's complex aviation system through transparency and sharing of information.[43]

8.6 Reflected Criticism

Notwithstanding the immense effort the European Commission put in establishing a EU blacklist, which so far, presumably, enhanced aviation safety within the European Union with the primary objective of protecting the interests of European citizens living in the vicinity of Community airports or travelling to non-EU destinations or travelling on non-EU air carriers, some questions are raised concerning the fact that blacklisting of individual air carriers may not be an effective and adequate or legitimate approach, as it really seems to be.

Temporarily suspending operation activities, but even worse, banning the complete operation of individual air carriers or specific aircraft or even all aircraft registered in some specific (Sub-Saharan African and Asian) countries by associations such as the

43 Roberto Kobeh González, President of the ICAO Council, in his opening speech at the High Level Safety Conference on 29 March 2010 in the Assembly Hall of the Headquarters of ICAO in Montréal, Canada (*source*: ICAO News Centre).

European Union, is generally perceived as subjective and unilaterally considered inefficient in remedying the real underlying problems inherent to inadequate safety oversight by national civil aviation authorities.

Furthermore, the EU blacklist is, despite the professional and independent approach of EASA and the Air Safety Committee, subject to economic and political influences. In general, blacklists are susceptible to political intervention and repercussion, in fact retaliation measures.[44] In "Will blacklists enhance airline safety?" the author, Andreas Knorr,[45] mentioned two examples: "The Dutch supervisory authority imposed a temporary entry ban on the Turkish charter air carrier Onur Air on 12 May 2005 after two incidents (without any personal injuries) became publicly known. The German, French and Swiss supervisory authorities shortly afterwards followed suit. Hereupon, after massive protests the Turkish Government temporarily, and against effective international regulations, refused entry or landing rights for EU air carriers, which should fly out stranded holidaymakers as a substitute for Onur Air. The entry flight ban imposed on Onur Air was revoked shortly afterwards. More recently, a SN Brussels Airlines Airbus 330 aircraft was held at Kigali airport in Rwanda, allegedly for lack of adequate maintenance documents, after one Rwandian cargo operator had been put on Belgium's national blacklist".

Critical Appraisal:
- Political considerations override safety objective
- • No major operator of Airbus or Boeing fleets has been affected by EU's and FAA's blacklists.
- • FAA initially did not apply IASA rules to Russia, China, Nigeria, but cleared them with "minimally passing marks".
- • Most banned carriers have never served EU/USA, or at best are marginal niche players (ethnic traffic)
- Safety record of most of the blacklisted carriers is not significantly worse than global/regional average despite their operating in extreme adverse environments.
- No clear correlation exists between oversight activities and airline safety standards.
- Ramp checks useful, but are only spottily practiced and do not cover some crucial safety aspects, such as CRM.[46]

---

44 *See also*: B. Scherer, 'Der Fall Onur Air', *Frankfurter Allgemeine Zeitung*, 19 May 2005, No 114, p. R1.

45 Prof. Dr. Andreas Knorr, Chair of Economics (National and International Economic Policy), Deutsche Hochschule für Verwaltungswissenschaften, Speyer, Germany.

46 A. Knorr, *Will 'Blacklists' Enhance Airline Safety?*, German Aviation Research Society (GARS) Workshop, Amsterdam, 29 June 2006, at 9.

Knorr further cites a number of factors affecting revision of the EU blacklist, such as corruption, incompetence, design or certification flaws, manipulation of facts, data and statistics and, in addition, some exogenous factors, beyond the reasonable control of air carriers, such as a fiercely competitive aerospace industry reduced to two market leaders (Airbus and Boeing), the quality of air traffic control and governmental supervision in specific countries, adverse weather conditions, inadequate local navigation facilities and default of other safety-related aids, terrorist actions and hijacking attempts.

> Blacklists do not positively contribute to the improvement of aviation safety. Not only do they fail to provide an objective as well as reliable indicator of air carrier or country safety standards. Furthermore blacklists do not identify, let alone help to remedy the underlying causes of safety problems and are, as a result, no adequate policy instrument to enhance safety standards of individual air carriers and in specific countries, in particular many poor and ill-governed third world countries.
>
> In fact, especially the EU blacklist is simply a grossly misleading bureaucratic exercise – a mere placebo for a safety-conscious but uninformed public. Fatal and non-fatal air carrier accidents will inevitably continue to happen – both to blacklisted air carriers and those unlisted air carriers which are deemed "safe" by supervisory authorities, policymakers and the media.[47]

With respect to updating of the EU blacklist, the European Commission is in some instances using a blanket banning of all carriers domiciled in the same country.

Dr. Elija Chingosho, since 1 January 2011 Secretary General of the African Airlines Association (AFRAA), states: "Last month, the European Commission updated its infamous list of air carriers banned in the European Union, on a blacklist of all carriers that have been certified in Mozambique, in addition to the operations of Air Madagaskar for two specific Boeing 767 aircraft, which the European Commission states is because of significant safety deficiencies requiring decisive action in both cases".[48]

However, the safety record of LAM Mozambique Airline is impeccable. Since the air carrier was established in 1980, not a single major accident has occurred. According to Dr. Chingosho, the air carrier has worked hard and invested significant resources to enhance its safety level, which enabled it to acquire the IATA Safety Audit Certification in 2007,

47 Id., at 20, 21.

48 E. Chingosho, 'Europe Should Abolish Its Airline Blacklist', <www.arabiansupplychain.com>, 6 June 2011. *See also* <www.african-aviation.com>, 'AFRAA Responds to EU Ban of Mozambican Airlines from European Airspace', 2 May 2011.

later renewed in 2009, together with the IOSA Certification and ISO 9000 Certification. Nevertheless, these awards have not spared the air carrier from the EU blanket banning.

In this context Dr. Chingosho is making a comparison: "According to the Flight Safety Foundation, Air France as a major European air carrier, has had 23 major accidents (involving substantial damage to aircraft, serious or fatal injuries) since 1990, three of them with fatalities, and a total of 348 deaths". Mozambique Airlines, although a smaller air carrier with limited capacity, was not involved in any accident whatsoever since 1989.

Blanket banning raises the question of partiality in an after-all balanced economic air transport market, besides the fact that the African Airlines Association is utterly fails to understand how such banning could contribute to encouraging African air carriers to aspire to reach the same level of safety standards as generally practised in Europe.

The EU blacklist appears to give preferential treatment to EU carriers to swiftly fill the gap in the air transport business to African states that are subject to the ban, in order to attain the market share of banned African air carriers, and subsequently to generate higher yields. "The banning of an air carrier not only prohibits that air carrier from operating to the European Union but also impacts its ticket sales to other destinations, including on code shared routes as travel agents and other code share partners in the European Union are required by regulation at the time of sales or booking to notify passengers that the airline is blacklisted. Irrespective of what the European Commission public relations exercise of attempting to pass blanket banning as a solution to safety concerns is, it amounts to nothing more than a blunt instrument that constrains the development of a viable African air transport industry".

On the other hand, Dr. Chingosho is well aware of a backlog regarding safety standards and governmental supervision with respect to civil air transport in some African States. He reiterates in the above-mentioned article the call of AFRAA upon all African stakeholders: (…) "to address the serious oversight deficiencies and concerns in the States blacklisted and meaningfully engage with the European Union to establish a mutually acceptable, fair and transparent mechanism to address safety concerns in place of the unilateral blanket banning, which has so far not yielded any meaningful achievement in advancing safety in the continent".

IATA is targeting all African members to cooperate in enhancing aviation safety in the region. IATA's Director General and CEO Giovanni Bisignani, at a press briefing in Johannesburg on 5 May 2011, declared:

> Africa must improve safety, the industry's number one priority. Last year the African accident rate was 12 times the global average. IATA African members, who are all on the IOSA registry, performed better – but aviation must be safe everywhere, and for all air carriers. Africa cannot and should not be the exception.

> The recent extension of the EU blacklist to include some IOSA registered carriers shows the limitation of the EU's approach to safety. IATA is here to help, and IOSA is the flagship. It is tough but achievable: 22 sub-Saharan African air carriers are on the registry. Six African air carriers did not meet the IOSA deadline to keep IATA membership in 2009, but we are working to bring them back on board.
> To drive improvements in safety we also need governments to be committed, and we are working with governments across the region to promote the use of IOSA.
> I urge South Africa, as the region's largest economy, to join IATA in taking a leadership role in promoting safety throughout the continent. To do business with the rest of the world, Africa needs a safe aviation industry operating to global standards".[49]

Cooperation and synergies are the key issues of today's air transport operations. To be linked or to cooperate in Africa with other national (government-owned) air carriers, or to be part of global air carrier alliances, will provide the ability to compete effectively in the world aviation industry market.

African air carriers, working in partnership with others, are very well able to benefit from each other's operations, business cultures, maintenance methods, networks and safety structures; however, they still struggle with their common safety level image and consequently the EU ban.

Dr. Chingosho: "You know the air carrier industry is very competitive. Now, when you are very competitive, people look for tricks to put the other parties into competitive disadvantage. Unfortunately, the European Union is using the issue of safety, which we believe is wrong. You don't say your airspace is not safe, don't fly to the EU … and you are saying that to an air carrier in Swaziland, for instance".

He continues: "That air carrier in Swaziland is a small carrier and this carrier has no interest in flying to the European Union. It is operating as a small air carrier.

It will need to stop, change arm before reaching Europe. So the question is why would the European Union be banning this air carrier? I think the whole issue is just to create an impression that African air carriers are not safe. Now, by creating that impression, the passenger coming from Europe to Africa would rather be flying a European air carrier. (…). And more and more are coming to Africa, that is why they are putting that just to create a wrong impression about Africa".[50]

---

49 Mr. Giovanni Bisignani was Director General and Chief Executive Officer of IATA from 2002 till July 2011.

50 E. Chingosho, 'Lack of Cooperation Killing African Carriers', *NBF News*, 17 October 2011.

8.7 SAFA AND THE EU BLACKLIST

SAFA is a spot examination regarding the operational status of the aircraft and its crew to be assessed. EASA, but also commercial organizations in the air transport industry, emphasize that an on-the-spot aircraft ramp inspection is limited in scope and nature, characterized by restricted time and checklist items, and as such should not be considered a proven fact or a fait accompli that the air carrier concerned is performing below standard regarding safety aspects and operational procedures, and consequently a victim for possible inclusion in the EU blacklist. As SAFA inspections are not very elaborate, they can only be indicative and the SAFA programme is certainly not intended to replace or take over the respective safety oversight responsibilities of the State of Registry or the State of the Operator.

Nevertheless, it should be emphasized that safety assessments of aircraft and aviation safety oversight audits on authorities generally contribute to the safe operation of aircraft, as operators and the authorities in their own interests usually take prompt measures to correct identified deficiencies.

Yet, there are a number of questions and ambiguities concerning how air carriers may or may not be placed on the EU blacklist. It seems that the question of which air carriers end up on the blacklist is determined less by safety concerns than by economic issues, such as networks, business connections and non-transparent political interests.

> Yemenia Airways also offers up an example of just how divergent the views of the European Union's various safety authorities are. The European Commission, where decisions are made by officials from EU member States, declared itself satisfied with Yemenia's maintenance. At the same time, EASA, the European safety agency, reached a very different conclusion, finding serious and systematic deficiencies at Yemenia's maintenance company.
> EASA, which is based in Cologne, revoked the Yemeni company's license for servicing EU planes on Feb. 24, 2009.[51]

Global aviation organizations such as ICAO and IATA are not in favour of the European system of blacklisting unsafe air carriers that do not comply with international safety standards. These organizations have the diplomatic tendency to prefer either the U.S. method of banning all air carriers registered in States whose civil aviation authorities do not enforce international safety standards, or focusing on individual air carriers by means of safety audits and to extensively assist in a correction process prior to any adverse action.

---

51 H.J. Schlamp, S. Simons, H. Stark & G. Traufetter, 'Flying Coffins. How Strict Is the EU's Airline Blacklist?', *Der Spiegel*, 14 July 2009.

The U.S. method is to focus on the ability of a country, not an individual air carrier, to adhere to international safety standards and recommended practices for aircraft operations and maintenance established by ICAO. It should be noted that the safety assessment by the FAA, being different from the European way of directly blacklisting unsafe individual air carriers, is considered rather arrogant and embarrassing, however fully lawful.

The actual and prolonged effect on global aviation safety, however, seems to be limited. Meanwhile ICAO (USOAP) and IATA (IOSA) do have their own safety audit programmes, the first primarily focusing on governmental safety oversight performance and the latter on individual air carrier (non)compliance with ICAO safety standards.

Both initiatives will thus have an influence upon safety performance; however, where USOAP may be considered a top-down approach, IOSA is solely working inside the structure of air carriers and, as such, is considered to have a bottom-up approach.

Although the principles of safety assessments by these organizations are different, they both intend to collaborate in order to enhance global aviation safety. In this respect information from ICAO plays a major role in the EU's blacklist policy. In July 2009, the European Union blacklisted nearly all of Kazakhstan's air carriers (71 in total) following a recent ICAO investigation that found serious problems.[52]

More recently, the European Union blacklisted air carriers because of significant shortcomings in the safety oversight system owing to inefficient and lax action of the authorities: "The EU's most recent aviation blacklist (8 January 2010) bars all of Kyrgyzstan's airlines from entering the Union's airspace, and all but one from Kazakhstan. The blanket ban reflects EU concern at safety deficiencies in the oversight system. Reflecting their Soviet-era legacy, a combination of weak financial resources, outdated legislation and lax regulatory requirements has affected development of the aviation sector in these countries. To meet EU requirements, substantial modernization of the sector will be required".[53]

8.8 Concluding Remarks

Whilst some tragic commercial aircraft accidents in the 1990s, causing loss of life of EU citizens, prompted the European Commission to define an aviation safety improvement strategy with subsequent legislation, three consecutive fatal aircraft accidents in 2005, involving mostly EU citizens, urged the European Commission, in consultation with member States' aviation safety authorities, to establish regulations to be able to fully or partially ban air carriers found to be unsafe from operating within the European airspace.

52 J. Renkin, 'EU Extends Airline Blacklist', <www.europeanvoice.com>, 7 July 2009.

53 A. Walker, 'Inclusion on EU Aviation Blacklist Spurs Reform in Central Asia', Archive of European Integration, EUCAM Commentary No 9, 8 January 2010.

The blacklist, an abomination to a lot of foreign air carriers, was created mainly with a view to formulating an efficient and coherent approach to reinforce safety, to give air transport passengers the right to be informed of the identity of the air carrier actually operating the flight they have booked, which is an essential piece of information that has to be transparent and recent, and the reimbursement or transfer to another air carrier in case of an imposed cancellation or a flight booked with an air carrier added to the blacklist.

The primary intent of the Community list is to ensure a high level of protection for passengers from safety risks, and, in general, to protect citizens on the ground from increased danger. Each EU member State shall enforce, within its territory, the operating bans included in the Community list in respect of the air carriers that are subject to those bans. Member States of the European Union should lay down rules on penalties applicable to infringement of the provisions of Chapter III of Regulation (EC) No 2111/2005 and ensure that these penalties are applied. The penalties, which may be of a civil or administrative nature, should be effective, fair, proportionate and dissuasive.

In addition, Regulation (EC) No 2111/2005 should not preclude the EU member States from introducing a quality labelling system for air carriers at the national level, for which the criteria might include considerations other than minimum safety requirements, in accordance with Community law.

Air carriers are placed on the Community list if evidence shows that these carriers are persistently and incorrigibly underperforming with respect to safety matters.

Findings of SAFA inspections, carried out at Community airports, reveal primary indications of non-compliance with internationally adopted (minimum) safety requirements. The inability to take corrective actions shall quickly be a condition for grounding the aircraft concerned. Poor aviation safety oversight, caused by deficient implementation of internationally agreed legislation, non-compliance with safety standards and lack of human and financial resources, are highly correlated with accident rates and low reliability of relevant air carriers.

The greatest threat to civil aviation safety in Africa is poor regulatory oversight, as identified by IATA, followed by inadequate safety management systems. ICAO USOAP audits in Africa reveal deficient safety implementations. To address these shortcomings, centralized civil aviation authorities in different regions, assisted by development programmes realized by technologically and economically most advanced States and regional or global organization projects (ICAO COSCAPS), have to be established to supplement the existing civil aviation authorities in the regional member countries by providing useful resources in a central pool system. On the other hand, African criticism points in the direction of these highly developed States, notwithstanding meaningful and constructive assistance from partnerships such as the European Union, that blacklisting African air carriers might be seen as subjective and inconceivable in the context of that very assistance.

By banning African air carriers, major European carriers are unbalancing the regional economic air transport market by usurpation, leaving the loose ends to African air carriers, which are degraded to second-class operators, feeling their restricted and crippling operation like a stranglehold. This imbalance might fuel trade barriers and more specific repercussions like arbitrary (ground for rescinding an agreement) and suspending conditions in bilateral air transport agreements.

Air carrier deficiencies, appearing from aircraft accident and serious incident investigation reports, are, aside from the above-mentioned findings, sources of information that in any form will contribute to a decision-making process regarding inclusion in the Community list.

On the other hand, conclusive and objective evidence concerning rehabilitation regarding full adherence to safety requirements, in other words that banned air carriers have to demonstrate the successful completion of remedial and corrective action at any level necessary in the field of flight operations, maintenance and management, will be a justification for withdrawal of the Community list by the European Commission after advisory considerations of the Air Safety Committee.

Continuous and close monitoring by civil aviation authorities is a prerequisite for ongoing and adequate compliance with safety requirements by rehabilitated air carriers. In this respect the regional COSCAP programmes could be of great help.

Regarding the European method of banning air carriers within their airspace, ICAO and IATA have divergent opinions. These global organizations prefer the U.S. governmental safety assessment system, which categorizes States (and their aviation safety oversight authorities) whose air carriers want to initiate or maintain air transport services to the U.S., rather than the unilateral and coercive banning of third-country air carriers from European airspace. IATA, in particular, is focusing on individual air carriers to assist in operational, financial and technical matters. The fundamental objective of IATA is to encourage the development of air traffic by promoting safe, regular and economic transport. As safety is the priority issue, IATA is providing for, among other things, essential professional support to the aviation industry by improving perceptions on the role of aviation in economic and global competitiveness, by assessing air carrier performance, by developing efficient training, by challenging unreasonable rules and charges and by striving for sensible and constructive civil aviation-related regulations.

Although the maxims of leading safety assessment programmes as well as their effects in the air transport world are different, the general trend of the efforts and results indicates an expression of the ultimate commitment to the aviation industry and the pursuit of enhancing global aviation safety.

9 Aviation Safety Information and Jurisprudence

9.1 Introduction

Aviation safety information is of paramount importance in identifying deficiencies representing a threat to civil aviation safety and in subsequently being able to enhance aviation safety on a worldwide scale by exchanging reports of safety occurrences. In addition, responsible aviation authorities and organizations should not only monitor and enforce compliance with safety standards, but should also promote uninhibited reporting and open exchange of all available safety information. To continuously mitigate risks, open exchange of aviation safety information is an indispensable and necessary tool, especially in the proactive phase.

Collecting, analysing, disseminating and sharing of information will ensure the acquisition of sufficient evidence for taking the necessary measures to guarantee the safety of air passengers, and residents living near airports.

Aviation safety information, in the broadest sense, is used as a communication system to be able to prevent recurrence of accidents by prematurely identifying and addressing certain shortcomings in the aviation industry. Not only is the issue of reducing and preferably eliminating poor maintenance, handling and substandard air operations of serious concern, but the extraordinarily rapid developments in aviation technology, advanced aircraft design, state of the art automation and high-tech control systems are demanding an incredible effort to adapt to complex future aspects and to systematically approach inherent interactions.

This requires all participants in the aviation industry to exchange information concerning technical developments, safety reports, practical experiences, human factors and the effects of applicable legislative rules and regulations.

Two mainstreams of creating information exchange systems are present in civil aviation. On the one hand, air carriers are collecting and exchanging safety information creating a combined database for analyzis and use as a benchmark for further refinement of the safety culture. On the other hand, governmental aviation safety organizations together with civil aviation non-governmental organizations agreed to strengthen global exchange of safety data to take more effective action to reduce risks and improve safety performance. During the ICAO HLSC in Montreal (2010), IATA, together with ICAO and the U.S. Federal Aviation Administration, and the European Commission, have all signed a Declaration of Intent to create a project regarding the global exchange of aviation safety data.

Audit data from various audit programmes that collect complementary safety information for analyzing and trend monitoring will be a key element in the project. "Today's milestone agreement marks the first time the global aviation community has come together to work on a global safety information exchange. Data must drive our actions so that we can focus our joint efforts on reducing the greatest risks."[1]

The system of safety information exchange, extended to be able to analyze combined databases of air operators, is used not only by civil aviation authorities and organizations but also by other stakeholders such as air operators.

The Safety Information Exchange (SIE), promoted by British Airways in the 1990s as an extension of their safety information management system BASIS, is in the process of making available (standardized, formatted) safety information, collected by various related organizations, to others with the aim of promoting safety and reducing accident rates.[2] IATA renewed this particular information exchange system and renamed it STEADES (Safety Trend Evaluation, Analysis & Data Exchange System).

This comprehensive system, in fact the world's largest database of de-identified incident reports, provides for a secure environment for participating air carriers to pool safety information for global benchmarking and analysis need. The analyses result in trend reports, occasionally giving rise to potential concern and can be accessed and used to develop an extensive list of prevention strategies applicable to a wide area of professionals within the air transport industry. IATA analyses all exchanged data for trends and issues of concern to identify prevention strategies. Generated trend reports are distributed to participating entities following the analysis process. The reports are used by air carriers to benchmark their safety performance.

The bottom line of the analyzing process and reporting is to contribute to global safety performance improvement. STEADES is one of more sources that provide IATA with global safety data through its Global Safety Information Center (GSIC), accessible to IATA members and other authorized users.[3]

In the European Union, EASA launched ESSI (European Strategic Safety Initiative) as a ten-year programme (2006-2016) aimed at promoting the highest common standards of safety in civil aviation and improving aviation safety in Europe, and for the European citizens worldwide. ESSI is an aviation safety partnership between EASA, other regulators, organizations and participants from the aviation industry such as aircraft and components manufacturers, civil aviation authorities, air carriers and pilot associations from

1 Speech during the ICAO High-Level Safety Conference in Montreal by Giovanni Bisignani, IATA's Director General and CEO (2010).

2 BASIS: British Airways Safety Information System is a two-way safety communications system that includes a telephone hot line, where pilots can report safety issues, and a computer-based system for distributing safety-related material throughout the airline. This system involves all employees, including pilots' reports of irregular crew operations. It forms the backbone of the programme's safety information collection and dissemination efforts (*source*: A.T. Wells & C.C. Rodrigues, *Commercial Aviation Safety*, 4th edn, McGraw-Hill Professional, New York, NY 2004, at 355).

3 The data sources of IATA GSIC are: IATA Safety Reports, IATA Operational Safety Audit (IOSA), IATA Safety Audit for Ground Operations (ISAGO), IATA STEADES and Flight Data Analysis Exchange (FDX).

EU member States, ECAC, EUROCONTROL, FAA, IATA and ICAO (more than 150 aviation-related organizations involved).

With regard to EC SAFA inspections it is regulated that information gathered in each member State shall be made available to all other member States and the European Commission to ensure the most efficient monitoring of compliance of third-country aircraft with international safety standards and thus establishing a satisfying, preferably high level of aviation safety.[4]

In the Community, but in fact throughout the world, exchange of information and access to related documents must be possible, particularly when such information shows the existence of a potential safety threat.[5] However, certain documents and information should be given special treatment because of their highly sensitive content. In this context, ICAO has initiated a major step forward by a particular agreement on this issue, adopted by 153 of 193 contracting States. This agreement ensures that by 23 March 2008 the names of all contracting States that did not agree to full transparency regarding the results of an USOAP review regarding their national aviation safety oversight system will be posted on the publicly accessible ICAO website. This information enables contracting States to obtain an objective overview of a particular contracting State's safety oversight system techniques and capability and of a specific air carrier's performance and safety record.[6]

States that have agreed to release information to the ICAO FSIX website are grouped into COSCAPs, based on region and further divided into organizations of specific responsibility. The premise is that cooperation between contracting States and exchange of information are essential elements for the success of any aviation safety-related activity, in pursuit of the common intention of enhancing aviation safety.

Information originated from SAFA inspections is considered to provide a general indication regarding the safety status of foreign operators. However, owing to the fact that certain aspects are difficult to assess during a rather brief ramp inspection, this indication can hardly give a complete impression concerning the safety of any particular aircraft or operator. Ultimately, more relevant information is in fact required.

9.2 JURISPRUDENCE IN THE NETHERLANDS WITH RESPECT TO SAFA INSPECTIONS AND EU BLACKLIST INCLUSION

Every decision to impose a ban must be the result of a conscientious and careful safety assessment and include a suitable right of defence for all air carriers under investigation. Notwithstanding the fact of profundity, the imposition of a ban can be done fairly swiftly

---

4 Directive 2004/36/EC under (9).

5 Regulation (EC) No 1049/2001 of the European Parliament and of the Council of 30 May 2001 regarding public access to European Parliament, Council and Commission documents, *Official Journal of the European Communities L* 145/43, 31 May 2001.

6 ICAO Flight Safety Information Exchange.

in urgent cases where an immediate threat to air safety is detected, either by ramp inspections, safety audits or through any other evidence providing method. Acting on a proposal of the European Commission, being chairman of the Air Safety Committee, though without any voting rights, the Committee adopts its opinion by qualified majority, which is then submitted to the European Parliament before final adoption by the European Commission and subsequent publication in the Official Journal of the European Union.

If the European Commission or a member State acquires evidence indicating serious safety deficiencies, consecutive deferred defects or malfunctioning (or the lack of ability or willingness to address these safety deficiencies) on the part of an air carrier or its responsible oversight authority anywhere in the world, they ask for the EU blacklist to be revised instantly. Member States have the obligation to ask for the update of this blacklist, followed by a decision based on the common safety criteria for banning established by the Basic Regulation (Regulation (EC) No 2111/2005). But initially, the civil aviation authority of a member State shall take, at the national level and time permitting, all appropriate measures to satisfactorily resolve any deficient operation.

Where the European Commission is of the opinion that the continuation of operations into or within the Community of an air carrier is likely to constitute a serious risk to flight safety and that such imminent risk has not been resolved satisfactorily at the national level, it can take provisional measures, whereby this particular air carrier is banned from entering European airspace. These specific measures are then presented to the Air Safety Committee for confirmation or potential modification.

At the national level, especially in but not limited to the Netherlands, poor performance by air carriers in the field of aviation safety, revealed by SAFA inspections (and otherwise obtained information), has led to lawsuits regarding alleged controversial suspension of air service traffic rights, in this case by the Dutch Government.

9.3 Phuket Airlines

Since 27 March 2005, Phuket Airlines Co. Ltd., of Thailand, was operating a non-scheduled air service between Bangkok and the Netherlands.[7] This international air service was authorized under the bilateral agreement (hereinafter referred to as the "Agreement") between Thailand and the Netherlands for air services between and beyond their respective territories (with route schedule).[8]

7 Phuket Airlines Co. Ltd. (Phuket Air) is a well-established air carrier for short- to long-term leasing and charter services. It is being organized to take advantage of a specific gap in the short- to long-haul domestic-international and regional travel market in specific areas around the world under a wet-lease-ACMI structure (<www.phuketairlines.com>).

8 Agreement Between the Government of the Kingdom of Thailand and the Government of the Kingdom of the Netherlands for Air Services Between and Beyond Their Respective Territories. Signed at Bangkok on 29 April 1971 (Registered by the International Civil Aviation Organization on 14 November 1975).

The reason for the immediate suspension of the entry permit of Phuket Air into Dutch airspace was as follows: Since 27 March 2005, Phuket Air was maintaining an air service on the route Bangkok-Amsterdam and vice versa with an aircraft of the type Boeing 747-306(M) (registration number HS-VAC). On 15 April 2005, this aircraft was subjected to a SAFA-inspection shortly after arrival at Schiphol international airport. This inspection revealed a total of six shortcomings: three deficiencies of category 3 (serious deviations), two of category 2 and one of category 1. One of the serious shortcomings consisted of not updated Jeppesen route, approach and departure charts for more than one year.[9] Normally these charts should be amended on a weekly basis. Provided that in the future this particular navigation equipment would be updated accurately, permission was given for the return flight to Bangkok.

However, a subsequent inspection at Schiphol on 17 April 2005 revealed no remedial action by the air carrier. Under these circumstances, repeated violation of the most serious category, the Dutch civil aviation authorities have been rightly deemed competent to impose an immediate temporary suspension, pending conclusive response of the Thai civil aviation authorities (committing to take corrective action and effective measures), in order to prevent consecutive violations of relevant laws and regulations.[10]

> Origin and course of the proceedings.[11]
>
> By decision of 19 April 2005 the defendant (Dutch State Secretary to the Minister of Transport and Water Management, hereinafter referred to as the "Dutch State Secretary")[12] has temporarily suspended the permit to use Dutch airspace, granted to the plaintiff (Phuket Airlines Co. Ltd., based in Thailand), pending conclusive answers by the Thai civil aviation authorities to some case-related questions.
> By decision of 22 April 2005 the defendant has lifted this suspension with immediate effect following replies from the Thai authorities on the same day.
> By decision of 29 April 2005 the defendant has revoked this permit.

9 Jeppesen (also known as Jeppesen Sanderson, Inc.), an American subsidiary of the Boeing Company, is specialized in navigational information, operations management solutions and (electronic) flight training products and services.

10 The European Union blacklisted (Commission Regulation (EC) No 474/2006 of 22 March 2006) the air carrier, banning it from flying to Europe in 2006.

11 LJN BA 2577 Rechtbank 's-Gravenhage (The Hague Court), case number: AWB 05/8479 BESLU, 21 March 2007.

12 From 2010 also referred to as: The Parliamentary Under-Secretary of State, Department of Infrastructure and the Environment (including Transport).

> By letter, dated 23 May 2005, the plaintiff has objected to the decision of 19 April 2005.
> By letter, dated 3 June 2005, the plaintiff has objected to the decision of 29 April 2005.
> On August 24 a hearing took place concerning the objectives of the plaintiff.
> By decision of 18 October 2005, and posted on the same day, the defendant has declared the objectives of the plaintiff partly founded as well as partly unfounded but has maintained the disputed decisions, except that the plaintiff's permit to use Dutch airspace has been suspended rather than revoked.
> The plaintiff subsequently appealed against the decision of 18 October 2005 on 28 November 2005 by facsimile.
> The defendant has submitted all relevant documents and a written defence. Then both parties have submitted some additional data.
> The appeal was heard in a court session on 7 March 2007.

Important in this legal dispute is Article 11 (Applicability of air regulations) of the Chicago Convention (The Kingdom of the Netherlands and the Kingdom of Thailand are parties to the Chicago Convention, opened for signature at Chicago on the seventh day of December 1944).

In this legal dispute the following documents also apply: the Agreement (Air service agreement between the Government of the Kingdom of Thailand and the Government of the Kingdom of the Netherlands, signed on 29 April 1971) and the exchange of notes constituting an agreement amending the route schedule annexed to the above-mentioned Agreement (hereinafter referred to as the "Agreement") on 31 October and 13 November 1974, in Bangkok.

Under Article 3.1 of the Agreement, each contracting party shall have the right to designate in writing to the other contracting party one or more air carriers for the purpose of operating the agreed services on the specified route.

In October 2004 Phuket Air has been designated for this operation by the Thai civil aviation authorities.

Under Article 3.6 of the Agreement, each contracting party shall have the right to suspend the exercise by an air carrier of the privileges specified in paragraph 2 of Article 2 of the Agreement or to impose such conditions as it may deem necessary on the exercise by an air carrier of those privileges in any case where the air carrier fails to comply with the laws or regulations of the contracting party granting those privileges or otherwise fails to operate in accordance with the conditions prescribed in the Agreement; provided that, unless immediate suspension or imposition of conditions is essential to prevent further infringements of laws or regulations, this right shall be exercised only after consultation with the other contracting party.

Article 2.2 states: Subject to the provisions of the present Agreement, the airlines designated by each contracting party shall enjoy, while operating an agreed service on a specified route, the following privileges:

a. To fly without landing across the territory of the other contracting party;
b. To make stops in the said territory for non-traffic purposes; and
c. To make stops in the said territory at the points specified for that route in the schedule to the present Agreement for the purpose of putting down and taking on international traffic in passengers, cargo and/or mail coming from or destined for other points so specified.

Crucial but obviously in this legal dispute is the deliberation regarding the acceptable level of safety, which is indicated by Dutch law concerning aviation safety, applicable in the Dutch jurisdiction. Article 3.1 of the Dutch Aviation Act of 18 June 1992 states:

An air carrier is obliged to ensure that:

a. aircraft operating under its operating license shall be in such a condition that transportation will be conducted safely;
b. flight crew members operating on board of its aircraft must have adequate knowledge, skill and experience;
c. That everything possible will be accomplished within its power to prevent serious physical or mental fatigue of operating flight crew members.

Under Article 5.3 of the Dutch Aviation Act it is prohibited to participate in air transportation or to provide air traffic control in such a way that persons or property are or may be endangered.

The legal dispute focused in the first place on the question whether the defendant deserves the right to suspend the granted permit of the plaintiff to make use of Dutch airspace.

Despite the argument of the plaintiff, the question whether the defendant deserves the right to revoke the granted permit is not addressed. Indeed, after reconsideration of the revocation decision of 29 April 2005, the defendant has concluded not to have the authority to proceed to revoke, but in reconsideration has taken the position that this virtual revocation is to be regarded as a suspension.

The court agrees with the defendant's conclusion, following the reconsideration process, because the plaintiff has not been disadvantaged thereby. Moreover, the court does not follow the plaintiff's view that disadvantage would lie in the fact that the plaintiff would have had to terminate its contracts with tour operators because of the revocation decision. After all, in case of suspension, the plaintiff would have been unable to meet its obligations to the tour operators to the same extent as in a revocation. As a matter of fact, the following aspect also applied, both in case of suspension and revocation, namely that the plaintiff would be eligible to again obtain the air service permit concerned when it became clear that the plaintiff would return to meet the standards.

It is assumed that the Dutch Government in principle derives its jurisdiction to prohibit aircraft to enter Dutch airspace from its sovereignty over its territory and corresponding airspace above.

This competence must be applied with due consideration to international law and may consequently be restricted by concluding treaties. The defendant has rightly judged that he deserves this competence, which originated from international law, because of the fact that this specific competence is consistent with the powers attributed by the Dutch Aviation Act.

The defendant's position is correct with respect to his competence to prohibit aircraft to make use of Dutch airspace, when it comes to the fact that the Agreement restricts this competence in that he is not authorized to revoke but to suspend the privilege to use Dutch airspace.

The Agreement therefore allows the defendant under Dutch sovereignty to suspend the entry permit into Dutch airspace when an air carrier fails to comply with Dutch laws and regulations. The court does not consider it wrong that the defendant, in order to specify the provisions of Articles 1.3 and 5.3 of the Dutch Aviation Act, has applied the standards which have been agreed by civil aviation authorities under the European Civil Aviation Conference and laid down in the SAFA programme (Directive 2004/36/EC of 21 April 2004).

With respect to the substantive grounds of appeal, the court considers the following: The defendant has based the first temporary suspension on the findings of the inspections on 15 and 17 April 2005. After the lifting of the temporary suspension on 25 April 2005, following defined promised measures by the Thai civil aviation authorities, inspections are again performed at Schiphol international airport. On 24 April 2005 two deficiencies of category 1 (the least serious) and one deficiency of category 2 were found. An inspection on 27 April 2005 revealed six deficiencies of category 1 and two of category 2. On 29 April 2005 three deficiencies of category 1 and two of category 3 were detected. However one category 3 deficiency was rehabilitated because after reconsideration it became clear that a medical certificate of the co-pilots' medical examination was valid for a 12-month period rather than 6 month.

To summarize: the basis for the first suspension was the fact that immediate action was needed following a repeated category 3 violation, and the basis for the second suspension was the fact that five inspections in a one-month period revealed a total of 23 deficiencies including 5 of the most serious category.

Regarding the manner in which the defendant has used its competence to suspend, the plaintiff argued that this procedural conduct is inconsistent with the principle of proportionality. The number and nature of identified deficiencies do not justify any suspension whatsoever. Moreover, the suspension that has been imposed is inconsistent, because the aircraft was released for departure only after minor repairs.

In addition, the plaintiff argued that the basis regarding the suspension decision of 29 April 2005 deemed to be unperceptive, because – unlike a statement in the decision – it appeared that the Thai civil aviation authorities had not proceeded to suspend the landing rights of the plaintiff.

The competence of the defendant to suspend the entry permit is a discretionary power. It is settled case law (jurisprudence) that the way the defendant has exercised this competence shall always be assessed by the court with restraint.

Regarding the claim of the plaintiff that the basis of the initial decision of 29 April 2005 has been dropped, the court considers the basis regarding the present disputed decision, basically the fact that the Thai civil aviation authorities would have suspended the landing rights of the plaintiff, to be unperceptive. The second suspension is yet based on the detected deficiencies as a result of performed SAFA inspections. The court considers this amendment of the basis not illegal. Indeed, the defendant assumed, based on a telephone communication with the Thai civil aviation authorities in accordance with the inspection on 29 April 2005, that the Thai authorities would suspend the landing rights due to the inspections performed in the Netherlands.

However, now it turned out that the Thai civil aviation authorities had not been proceeded to suspend, the defendant was to determine the findings of the SAFA inspections as a basis of the suspension.

In the opinion of the court, the defendant could reasonably come to the conclusion, in weighing all relevant interests, that the observed repeated violations of the third category on 17 April 2005 should lead to immediate temporary suspension for the duration of actions to be taken by the Thai authorities.

Furthermore, the court is of the opinion that the defendant has to come to the conclusion, based on the findings of the inspections on 15, 17, 24, 27 and 29 April 2005, that there was a pattern of negligence and careless dealing with respect to safety standards in daily operations. On that basis, the defendant has reasonably made its decision, in weighing the interests involved, to suspend the plaintiff's entry permit into Dutch airspace.

When weighing these interests, the defendant rightly invoked the fact that he could take no risks when it comes to safety aspects regarding air traffic. In other words, safety interests relating to air traffic are extremely important. Although some importance is attributable to the (financial) interests of the plaintiff with respect to this present decision process, these interests must however be considered as subordinate.

The court considers it not unreasonable that the defendant has decided to suspend, despite the fact that only two of the SAFA-forms (namely the forms of 17 and 29 April 2005) mention such a deficiency (category 3) that entry repercussions should be imposed. Plausible is, as the defendant has stated, that the inspector did not independently reach the conclusion that a category 3 deficiency should result in a suspension of the entry permit. It is common practice that the inspector only will allocate this serious qualification (3d) after consultation at a higher level, whereby the measures to be taken can be observed in a wider perspective and where the frequency of identified deficiencies may play a decisive role. This is confirmed by the fact that following the inspections of 17 and 29 April 2005, entry permit repercussions have taken place in the form of suspensions.

The court ruled, after having taken note of Appendix 4 of the SAFA programme on request of the plaintiff, that the defendant did not consider the magnitude of the findings of the inspections as too extreme, nor has associated disproportionate consequences.

Finally, the plaintiff has argued that the defendant, if not contrary to the principle of equality, at least acted arbitrarily and negligently, because in 2004 and 2005 a large number of categories 2 and 3 findings have been observed during inspections on another air carrier, Onur Air, which violations have not (immediately) led to suspension of an entry permit.

With respect to that particular case, the defendant stated in the first place that he eventually proceeded to suspend the entry permit of Onur Air. Then he declared that the reason why no intervention has taken place at an earlier stage is entirely related to the differences between the two air carriers, especially the way they have primarily responded on the inspection findings. In the case of Onur, to a greater extent than the plaintiff, ups and downs were present in its operations, leading to a later intervention. The court ruled that the defendant has made sufficiently clear that no similar cases are at issue and that the differences in circumstances justify the fact that actions taken against the air carriers in question are not exactly the same. Now the defendant also proceeded to suspend the entry permit of Onur Air, there is no plausible reason for violating the prohibition of arbitrary.

The court concluded, in weighing the interests involved, that the defendant could have reasonably come to the decision to suspend the Dutch airspace entry permit granted to the plaintiff, Phuket Airlines.

With regard to the ruling of the court in case No AWB 05/8479 on 21 March 2007, Phuket Airlines has appealed to a higher court on 4 May 2007.[13]

Appellant, Phuket Airlines, challenges the opinion of the court that the Dutch State Secretary was authorized to suspend the entry permit into Dutch airspace of Phuket Airlines.[14] Referring to the ruling of the Department of Administrative Law of 10 April 1995 (AB 1995, No 498; Long Lin) and the decision of the High Council of 7 February 1986 (NJ 1986, No 477; Attica Unity) appellant points out that although it is recognized in the relevant jurisprudence that the Dutch Government under certain conditions can derive its

13 LJN BC 3585 Raad van State (Dutch Council of State), case number: 200703161/1, 6 February 2008.

14 The Dutch governmental system recognizes various departments, each headed by a Minister and one (or more) State Secretary to the Minister.

power from the territorial sovereignty (including the corresponding airspace) to prohibit ships and aircraft to enter Dutch territory (territorial waters and land) or airspace, but in the aforementioned cases, according to the appellant, there was the complete absence of national legislation under which the Dutch Government could act. In this case, the Dutch Aviation Act, however, provides the authority under which actions may be taken against violations.

In proceeding to suspend the entry permit under different rules, the Dutch State Secretary thwarted the system of the Dutch Aviation Act. Appellant points out that the Agreement does not provide for direct competencies, and furthermore that, as far as it may be possible, the Dutch Government, rather than just the Dutch State Secretary, should decide to suspend. The decisions to suspend derive from Article 3, paragraph 6 of the Agreement.

The exercise of the right of the State of the Netherlands, as laid down in that treaty line, *in casu* to suspend under defined conditions the entry permit of an air carrier assigned by the Thai Government to operate that particular route, belongs in this particular case to the Dutch State Secretary.

To that end the court has correctly considered that the competency to suspend actually matches the competencies that the Dutch Aviation Act has attributed to the Dutch Minister of Transport.

In this context one can particularly refer to the authority to apply administrative measures to enforce the obligations imposed by or pursuant to the Dutch Aviation Act. Therefore, using this competency does not lead to illegally thwarting the system of the Dutch Aviation Act. Based on the portfolio allocation between the Dutch Minister of Transport and the Dutch State Secretary, this competency was exercised by the latter at the relevant time. Insofar the argument fails.

Appellant challenges the finding of the court, that it is not incorrect that the Dutch State Secretary, for further interpretation of Articles 1.3 and 5.3 of the Dutch Aviation Act, has applied the standards, agreed between civil aviation authorities in the context of the European Civil Aviation Conference, and as reflected in the SAFA inspection programme. According to appellant, the decision on appeal gives no indication that the standards of the Dutch Aviation Act are further completed by the SAFA inspection programme.

Appellant argues that the court failed to recognize the fact that the decision on appeal has been insufficiently motivated to that extent.

The International Civil Aviation Organization has established safety standards. Compliance of these standards by visiting foreign air carriers in the Netherlands will be checked by applying the SAFA inspection system. In the decision on appeal is listed that during these inspections a list of most common deficiencies is to be used. These deficiencies are, according to the nature of the effect on aviation safety, classified into three categories, whereby findings of category 1 will have a minor impact, of category 2 a significant impact and of category 3 a major impact on aviation safety. Findings not categorized will be assessed by the inspector with the list as comparison material. The decision on appeal also mentions certain actions to be taken dependent on the applicable category. A finding of category 3 may lead to denial of further entry permission.

The Dutch State Secretary considers, according to the decision on appeal, that appellant has violated Articles 1.3 and 5.3 of the Dutch Aviation Act. That decision, as to the seriousness of the observed deficiencies, was referring to the category classification according to the SAFA methodology. From this category reference can be inferred that the Dutch State Secretary has unmistakably applied the standards as laid down in the SAFA inspection programme, in order to assess whether the aforementioned provisions of the Dutch Aviation Act have been complied with.

There is no reason to call – and appellant has presented no relevant arguments – why the Dutch State Secretary, in all fairness, should not apply these standards. The decision on appeal, to that extent, is insufficiently motivated. Insofar the argumentation fails.

Appellant argues that the court wrongly considered that the Dutch State Secretary by decision of 19 April 2005, without prior consultation with the Thai civil aviation authorities, was allowed to proceed to immediate suspension, in order to prevent future violations of laws and regulations. According to appellant there was no immediate danger regarding flight safety. Furthermore appellant argues that, although with respect to the decision of 29 April 2005 prior consultation with the Thai authorities has taken place, the court however disregarded this consultation to be solution-focused, which according to appellant was not the case.

The Dutch State Secretary should observe international law when taking a decision. In this case it means that in principle before taking any primary decisions consultation with the Thai authorities must have taken place. In the decision making, which resulted in the actual decision of 29 April 2005, this requirement has been fulfilled. The court has rightly considered that appellant's statement, regarding the fact that this consultation was non-solution-focused, had no added value.

However, prior to the decision of 19 April 2005 no consultation has taken place, as mentioned before, because, according to the decision on appeal, the Dutch State Secretary felt compelled to act without hesitation, based on the pattern and seriousness of the findings on 15 and 17 April 2005. The deciding factor for this action was the fact that the inspection of 17 April 2005 revealed that no amendments or replacements have been carried out regarding the Jeppesen route, airport approach and departure charts. Missing of an updated Jeppesen navigation-set is considered a category three deficiency. The same deficiency was also detected two days before.

In all reasonableness, the Dutch State Secretary has been able to take the view that these conditions justified immediate action. The fact that appellant on 17 April 2005 still did not have a completely updated Jeppesen set comes at the expense and risk of appellant. Appellant knew, or at least should have known, that an updated version of a Jeppesen set on board of an aircraft is an essential requirement for safe operation. Insofar the argumentation fails.

Furthermore, appellant argues in vain the opinion of the court regarding the fact that the failure to give appellant an opportunity to express its view, *in casu* prior to the primary decisions, has been restored in the appeals procedure by hearing appellant in that procedure with respect to its view.

The argument of appellant that the court failed to recognize the fact that during the decision on appeal procedure the decision of revocation (29 April 2005) could not be converted into a decision to suspend, fails as well.

The position of appellant in fact is not deteriorated by the decision on appeal. Furthermore, there is no question of a new primary decision against which an appeal would be open, now in fact a permanent deprivation of rights is only converted into a temporary one. Where appellant is claiming to have suffered damage as a result of the initial decision of revocation, the Council Division considers that the decision on appeal does not involve a decision on a request of indemnification, and so the damage is not part of the subject matter of this dispute.

Contrary to the claim of appellant, the court has validly ruled that the Dutch State Secretary could modify the factual basis of the decision on appeal of 29 April 2005. This modification, alleging that the basis with respect to the decision on appeal will not imply the fact that the Thai authorities would suspend appellant's flights to the Netherlands, but whereas during consecutive ramp inspections deficiencies have been identified, has to be considered as a result of the review.

The Dutch State Secretary was in the false assumption, as appeared in appeal, that the Thai authorities would suspend the traffic rights based on findings during inspections performed in the Netherlands. By subsequently establishing these findings as a basis to the decision, the Dutch State Secretary did not exceed the limits of the review regarding the initial decision.

Appellant also argues that the court failed to recognize that the decisions retained to the decision on appeal were in fact disproportionate.

During a SAFA inspection on 15 April 2005 the following deficiencies, three of which are classified as category three, were identified:

1. Passengers remained on board during refuelling, while the cabin crew members were not in place at their (emergency) exit stations, in order to be able to assist in the rapid evacuation of passengers in the event of fire (category 3);
2. Jeppesen route and airport approach and departure charts were outdated (in some cases almost one year overdue (category 3));
3. The third navigation display was unserviceable; this deficiency was not entered in the appropriate list (category 3);
4. The Flight Operations Manual was faulty and the Basic Operations Manual was completely missing (category 2);

5. Several screws on the engine pylons no 2 and 3 were missing (category 2);
6. Engine no 4 was leaking fuel (category 1).

During the inspection on 17 April 2005, the Jeppesen charts were not yet replaced.

These findings and the emerging pattern of careless and negligent dealing with safety standards, could be and even were in fact, to the opinion of the Dutch State Secretary, a basis for temporary suspension of the traffic rights granted to appellant. Subsequent inspections (24 and 27 April 2005) revealed a number of deficiencies.

The Administrative Law Department of the Council endorses the opinion of the court that the Dutch State Secretary, in weighing all relevant interests, could reasonably come to the aforementioned decisions. With respect to this weighing of interests, the Dutch Minister of State has, as the court ruled, rightly implicated that he could not take any risk with respect to civil aviation safety aspects and that civil aviation safety interests should be treated with extreme caution.

Contrary to appellant's argumentation, the Dutch State Secretary was very well authorized to take into account the observed pattern of negligence as well as careless handling regarding safety standards during the operations.

The statement of appellant that on 17 April 2005 a completely amended Jeppesen set was not yet available, does not alter the fact that flight safety was an issue.

By operating that particular flight without the proper navigation set, appellant definitely did not comply with the safety standards. The fact that the aircraft after each inspection was released for departure, obviously after repair of identified deficiencies, does not necessarily mean that the safety condition of that aircraft was not an issue anymore. The argumentation insofar fails.

Finally, appellant argues that the court failed to recognize that the Dutch State Secretary has acted in opposition to the principle of equality, or acted arbitrary and carelessly, because for years he did not act against the Turkish air carrier Onur Air, while SAFA inspections on Onur Air revealed a large number of categories 2 and 3 deficiencies.

This argumentation also fails. The Dutch State Secretary has made it plausible that there was no reason to consider the different cases as being similar. Contrary to Onur Air, Phuket Airlines suffered from identified deficiencies with a more structural nature, almost immediately after granting the carrier its traffic rights, according to the Dutch State Secretary.

Onur Air, unlike appellant, operates flights to the Netherlands with multiple aircraft, and at that time was subject to a varying safety-quality level.

The appeal case is unfounded as follows from the above. The contested decision must be confirmed.[15]

---

15 Decision of the Administrative Law Department of the Dutch Council of State on 6 February 2008.

In 2006 the European Union blacklisted Phuket Air because of verified evidence of serious safety deficiencies. Commission Regulation (EC) No 474/2006 mentioned in paragraph 46 and beyond that:

> There is verified evidence of serious safety deficiencies on the part of Phuket Airlines. These deficiencies have been identified by member States, the United Kingdom and the Netherlands, during ramp inspections performed under the SAFA programme.[16]
> Phuket Airlines demonstrated a lack of ability to address timely and adequately these safety deficiencies.
> The authorities with responsibility for regular oversight of Thailand did not fully cooperate with the civil aviation authority of the Netherlands when concerns about the safety of Phuket Airlines certified in that state were raised as demonstrated by the lack of pertinent responses to the correspondence from this member State.
> Therefore, on the basis of the common criteria, it is assessed that Phuket Airlines does not meet the relevant safety standards.

Phuket Airlines commissioned Air France Industries, the aircraft maintenance division of French flag carrier Air France, to assist in raising the safety standards of Phuket Airlines, which was eventually successful in lifting the EU ban.[17]

9.4 ONUR AIR

Content indication Case no AWB 05/6303:

Authority to suspend the entry permit regarding non-scheduled air traffic flights.

The court (The Hague Court), jointly with the plaintiff, is of the opinion that from the provisions of Article 6 of the Dutch Resolution on Unscheduled Air Transport (*Besluit ongeregeld luchtvervoer BOL*) and the accompanying Explanatory Memorandum (together hereinafter referred to as "the Resolution") follows that the interest of aviation safety does not belong to the interests which this Resolution must protect. Economical assessments only play a role in granting landing rights in unscheduled air transport under this specific Resolution.

Consequently, revoking or suspending landing rights (or entry permits) can only be accomplished if based on (aviation-related) economical factors, stated in the Resolution. Appeal declared well-founded, and contested decision annulled.

---

16 CAA-UK-2005-40,-41,-42,-46,-47,-48. CAA-NL-2005-49,-51,-54,-55,-56.

17 B. Straus, *Europe Lifts Ban on Phuket Airlines, Adds PIA to Updated Blacklist*, Air Transport World, 6 March 2007.

The court sustains the legal effects. The court rules that the defendant, based on the sovereignty of the Dutch Government over its own territory, had the authority to proceed to suspend the landing rights because of safety reasons. The court is of the opinion that the defendant, by taking his decision to suspend, may rely on the findings of the relevant SAFA-inspections, unless it immediately appears that these findings are inaccurate.

The plaintiff argues in vain that the number of deficiencies regarding its aircraft does not deviate from average. The number tends to be even more positive than average. The court judged the position of the defendant as being not incorrect to the extent that the defendant could not be considered committed to firstly assess, in the event that the defendant identifies an alarming pattern of negligence and carelessness regarding safety standards in the operation of the respective air carrier, if other air carriers are demonstrating the same or similar alarming pattern, prior to intervene.

All in all it is not clear from the a posteriori evaluation expert reports, that the defendant has overestimated the seriousness of the problems regarding the safety standards in general, and in particular in the area of – weight and balance, wheels, tyres and brakes – as well as defect reporting and handling, which occurred in the plaintiff's daily operation.

In view of the foregoing, the defendant may have concluded that in the daily operation of the plaintiff, *in casu*, the dealing with safety requirements was at such a deplorable level, that any further permission to make use of Dutch airspace was not justified, and that the defendant, based on this information, should proceed to suspend the plaintiff's traffic rights.

Judgement in the lawsuit between the plaintiff, Onur Air Tasimacilik A.S. located in Istanbul, Turkey, and the defendant, the Dutch Minister (previously the Dutch State Secretary) of Transport and Water Management.

Origin and course of proceedings.

By decision of 12 May 2005, motivated on 17 May 2005, the defendant has provisionally suspended the traffic rights regarding unscheduled air transport of the plaintiff for a period of one month. The plaintiff has raised objections against this decision by letter of 18 May 2005.

By decision of 24 May 2005 the defendant has lifted the suspension by imposing certain conditions, namely adequate execution of the action plan established in the context of the Joint Aviation Authorities-meeting on 23 May 2005.

The plaintiff objected this conditional procedure by letter of 7 June 2005.

On 15 June 2005 the plaintiff attended a defendant commission-hearing.

By decision of 27 July 2005, sent at the same date, the defendant has declared the objection of 18 May 2005 well-founded, insofar the objection is directed against the motivation of the decision of 12 May 2005. However, the defendant has declared the remainder of the present objection unfounded, and additionally has declared the objection of 7 June 2005 against the decision of 24 May 2005 unfounded as well.

The plaintiff has lodged an appeal by letter of 5 September 2005 against this decision. By letter of 28 November 2005 the plaintiff has provided this appeal with motivations.

The plaintiff has further produced an expert report of 10 November 2005 of the International Bureau of Aviation in London (IBA).

The defendant has produced a written statement of defence with respect to the provided legal documents and report relevant to the lawsuit.

In a reaction to the production of the IBA report by the plaintiff, the defendant has produced a research report of the Dutch National Aerospace Laboratory (NLR *Nationaal Lucht- en Ruimtevaartlaboratorium*) by letter of 28 June 2006.

By letter of the same date the defendant has produced a further document regarding the lawsuit, viz. the Handbook of SAFA Ramp Inspection Procedures. The defendant has requested to restrict the cognizance of Appendix 4 to the court only, by invoking Article 8:29 paragraph 1 of the General Administrative Law Act.

By decision of 7 August 2006 the single judge of the court has rejected this request, because the plaintiff produced this current document at an earlier stage.

The plaintiff has requested the court to question the defendant in order to produce the following documents:

1. Correspondence and documents regarding the JAA-meeting on 23 May 2005;
2. The analysis report of the JAA SAFA ORB of 20 May 2005;
3. Correspondence between the Inspection Division of the Ministry of Transport and Water Management and foreign civil aviation authorities before and regarding the decision of 12 May 2005.

By letter of 6 February 2007 the defendant has produced these documents. Furthermore, the defendant has requested, by invoking Article 8:29 paragraph 1 of the General Administrative Law Act, to restrict the cognizance of these documents to the court only.

By decision of 8 February 2007 the single judge of the court has considered the restriction of the cognizance as legitimate, except for one document.

On request the plaintiff has given no permission to give judgment . This move was partly based on documents for which the court has deemed secrecy as being justified. Then the case was referred to another trial chamber.

By letter of 19 July 2007 the plaintiff further has laid the foundation regarding its appeal by presenting several productions, including a supplementary report of the IBA of 12 July 2007. The translation of this report has been submitted by letter of 16 August 2007.

By letter of 16 August 2007 the plaintiff responded to a number of points from the written defence.

By letter of 30 August 2007 the defendant responded to the plaintiff's letter of 19 July 2007 as well as to the supplementary IBA report.

The hearing of the appeal was on 20 September 2007.

Motivation

Article 11 (Applicability of air regulations) of the Chicago Convention (Turkey and the Netherlands are being contracting parties to this Convention) states:

> Subject to the provisions of this Convention, the laws and regulations of a contracting State relating to the admission to or departure from its territory of aircraft engaged in international air navigation, or to the operation and navigation of such aircraft while within its territory, shall be applied to the aircraft of all contracting States without distinction as to nationality, and shall be complied with by such aircraft upon entering or departing from or while within the territory of that State.

The Government of the Republic of Turkey and the Government of the Kingdom of the Netherlands, being parties to the Convention on International Civil Aviation and the International Air Services Transit Agreement, both opened for signature at Chicago on the seventh day of December, 1944, have signed the bilateral Air Transport Agreement (Trb. 1971, 199) at Ankara on 17 September 1971. This Agreement is related only to scheduled civil air transport. The Air Transport Agreement has been modified on 1 February 1983 at Ankara and on 10 June 1988 at The Hague by Memorandums of Understanding and Agreed Minutes done at Ankara in 1997.

According to the Agreed Minutes, meant as supplemental and amending agreement details, signed on 22 January 1997, an arrangement for unscheduled air transport has been established (part I), based on existing policy. Both parties agreed that the amendments should have immediate effect.

According to part D of the Agreed Minutes, both parties also agreed on the subject which is included in the Annex "Aviation Safety" (the Annex), as long as a standard aviation safety clause of ICAO (ICAO standard) is not available.

Part 2 of the Annex states:

> Either contracting party may request consultations concerning the safety standards maintained by the other contracting party relating to aeronautical facilities, aircrews, aircraft and operation of the designated airline. If, following such consultations, one contracting party finds that the other contracting party does not effectively maintain and administer safety standards and requirements in these areas that at least equal the minimum standards that may be established pursuant to the Convention, the other contracting party shall be notified of such findings and the steps considered necessary to conform with these minimum standards, and the other contracting party reserves the right to withhold, revoke, or limit the operating authorization or technical permission of an airline designated by the other contracting party in the event the other contracting party does not take such appropriate corrective action within reasonable time.

Pursuant to Article 1.3 of the Dutch Aviation Act, an air carrier is obliged to take care of the following:

a. That aircraft operating under the air carrier's operating license shall be in such condition that transportation will be conducted safely;
b. That the flight crew members on board of aircraft being operated by the air carrier must have adequate knowledge, skill and experience;
c. That everything possible will be accomplished within the air carrier's power to prevent serious physical or mental fatigue of operating flight crew members.

Under Article 5.3 of the Dutch Aviation Act it is prohibited to participate in air transportation or to provide air traffic control in such a way that persons or property are or may be endangered.

Article 76, paragraph 1, opening words and sub k, of the Dutch Aviation Act, provides, by virtue of an Order in Council, further instructions regarding the rules, which air carriers must comply with regarding the performance of particular types of air transportation within, to or from the Netherlands, or using the Netherlands as an intermediate stopover.

Pursuant to the paragraph of this Article 76, among other things, the Dutch Resolution on Unscheduled Air Transport (*Besluit ongeregeld Luchtvervoer* BOL, *Stb.* 1975, 227) has been established.

Pursuant to Article 3 of the BOL it is only permitted, unless otherwise defined by international agreement, to operate flights in unscheduled air transport by virtue of a specific authorization of the defendant directed to the air carrier concerned.

Pursuant to Article 6, paragraph I of the BOL, the following will be taken into account when considering the relevant decisions to be taken:

a. The interests of consumers in a coherent network of scheduled air services, which provides for sustainable and reasonable air transport requirements within, into or from the Netherlands or regarding the Netherlands as an intermediate stopover.
b. The interest of consumers with respect to unscheduled air transport, to suit their needs, at the lowest possible cost.
c. The importance of profitability of the air carriers. Pursuant to Article 6, paragraph II of the BOL, flights in unscheduled commercial air transport will be authorized if this kind of air transport, by its conditions of transport, apparently intends to meet other needs than the existing scheduled air services, and therefore it is plausible that it shall not affect any existing scheduled services.

Pursuant to Article 6, paragraph III of the BOL, the objective will be to prevent unsound competition. Unsound or unhealthy competition exists if the yields, produced by air carriers in a sound air transport operation, scheduled as well as non-scheduled, do not enable the operating air carriers to obtain a reasonable profitability in the long term. Pursuant to

Article 6, paragraph IV of the BOL, internationally agreed rules and commitments and the relations in the field of international aviation policy are taken into account.

Pursuant to Article 6, paragraph V of the BOL, in principle no distinction will be made between Dutch and foreign air carriers, regarding flights to or from the Netherlands or flights using the Netherlands as an intermediate stopover, provided that within the State where the foreign air carrier is domiciled, Dutch air carriers will have equal access to conduct flight operations in unscheduled air transport.

Primarily between the parties in this dispute is whether the defendant could derive the power of authorization from Article 3 of the BOL, to suspend the plaintiff's granted traffic rights to operate flights in non-scheduled air transport (route traffic right permit) based on reasons related to aviation safety.

The court, in conjunction with the plaintiff, is of the opinion that from the provisions of Article 6 of the BOL and the supplementary Explanatory Memorandum to the BOL, it is clear that the interest of aviation safety does not belong to the interests which the BOL intends to protect.

In granting traffic rights for flights in non-scheduled air transport based on the BOL, only (aviation)-economic interests play a definite role. Consequently, refusal or suspension of traffic right permits could only be carried out pursuant to (aviation)-economical factors as mentioned in the BOL. The provision in Article 6 paragraph IV may only relate, contrary to the defendant's argument, and in view of what has been considered, to internationally agreed rules and commitments which will ensure these (aviation)-economical interests.

The court comes to the conclusion that the defendant could not derive the authorization to suspend traffic rights granted to the plaintiff from the provisions of Article 6 of the BOL in combination with the Agreed Minutes of 22 January 1997, including the Annex, titled "Aviation Safety". The appeal should therefore be declared well-founded and the contested decision taken by the court should be annulled.

Then the question arises whether there is reason to maintain the legal effects of the annulled decision by the application of Article 8:72, paragraph 3 of the Dutch General Administrative Law Act (*Algemene wet bestuursrecht Awb*).

With respect to the power of authorization of the defendant in order to proceed to suspend the traffic right permit to enter Dutch airspace, it must be preconceived that the Government of the Netherlands in principle derives its power from the sovereignty over its territory including the corresponding airspace to prohibit aircraft from using Dutch airspace. This power has to be applied with due consideration regarding international law and therefore may be limited by the conclusion of relevant agreements.

The defendant has validly considered that this particular power of authorization, arising from international law, accrues him, because this power matches the powers of authorization which are attributed by the Dutch Aviation Act to the defendant, which in this case is either the nominated Minister or the State Secretary.

The defendant has rightly taken the view that neither the Chicago Convention nor the Air Transport Agreement between the Governments of Turkey and the Netherlands have limited the defendant's power of authorization to suspend air traffic right permits. The court follows the opinion of the plaintiff that the Supplement to the Agreed Minutes did not become part of the Air Transport Agreement according to the procedure mentioned in Article 12 of that Agreement. Moreover, the content of the Supplement does not affect the power of authorization of the defendant to suspend traffic right permits.

The court concludes that the defendant, based on the territorial sovereignty of the Government of the Netherlands, was authorized to suspend traffic right permits because of safety reasons. The plea of the plaintiff, that the suspension of a traffic right permit should be regarded as a punitive sanction, fails.

After all, the suspension of a traffic right permit does not aim at inflict suffering, but to restore a safe situation. The fact that it could lead to substantial damage to the party involved, does not lead to the conclusion that this damage was in fact envisaged. The seriousness of the recovery sanction itself is not such that on that particular basis already has to be considered that this sanction has a punitive nature, where the court is taking into account the fact that there is no long interval between the immediate cause of the suspension, the number of incidents in May 2005, and the suspension itself, and the fact that the suspension of the traffic right permit was lifted as soon as the defendant considered it justified on the basis of the Action Plan.

The court must also assess whether the defendant had reasonable grounds to decide to suspend the traffic right permit in this particular case.

It is established case-law that the court needs to reluctantly review the way the defendant has used its discretionary power of authorization.

The court does not consider it incorrectly that the defendant has applied the SAFA programme standards to further supplement the provisions of Articles 1.3 and 5.3 of the Dutch Aviation Act. These standards are agreed upon between civil aviation authorities in the context of the European Civil Aviation Conference and which are reflected in the SAFA inspection programme.

The defendant has based the decision to suspend the traffic right permit, as summarized, on the determination that in a short period before 12 May 2005 a number of incidents occurred with aircraft of the plaintiff, preceded by a period (early 2003 until 10 May 2005) in which timeframe a number of incidents and an accident occurred and during SAFA inspections repeatedly serious deficiencies of categories 2 and 3 have been detected. From all the findings and incidents as well as addressing these cases, a pattern emerged of disorganized and careless handling in the daily air operation with respect to safety standards.

According to the contested decision, the defendant considered it in particular important, in spite of the continuing dialogue with the plaintiff concerning the identified deficiencies, that no real improvement has been observed.

The sequence of incidents which occurred between 30 April 2005 and 10 May 2005 justified the conclusion that the plaintiff had learned no lessons from identified shortcomings of the previous period.

The plaintiff has disputed, referring to the findings of the IBA reports, that there was a deviation from the average in the negative sense with respect to the safety requirements and also regarding the dealing with the findings detected during the SAFA-inspections. Regarding the incidents, the plaintiff has stated, referring to the same reports, that such incidents could happen with any air carrier and actually will happen, and that such incidents, consequently, do not simply indicate a pattern of disorganized and careless handling in the daily operation with respect to safety requirements.

The court is of the opinion that the defendant, in taking a decision to suspend traffic right permits, may rely on the findings of SAFA inspections, as laid down in the SAFA Ramp Inspection Reports (hereinafter referred to as "Inspection Forms"), unless the inaccuracy of these findings is immediately apparent. The obligation to carefully prepare a decision to suspend is not going so far that the defendant is required to check, by means of an expert report, if each detected deficiency properly has been accounted for under the SAFA standards. In this respect the court takes into account that the submitted expert reports indicate that, regarding the technical aspects of a finding and its implications on safety, disagreement is possible at a detailed level.

In this context the court determines that there is no evidence that the plaintiff, prior to the suspension decision, has expressed that, to the plaintiff's opinion, the Inspection Forms contain wrongly classified observations (as findings) or contain incorrectly categorized findings. Not earlier than in appeal, the plaintiff has, referring to the IBA report of 10 November 2005, taken a position that of all the invoked findings only 13 were to be considered as noteworthy, of which only 1 finding was identified as category 2 (a worn brake unit), which however, according to the plaintiff, never has jeopardized flight safety.

From the produced IBA expert reports and the report of the Dutch Aerospace Laboratory it does not follow that the findings which according to the Inspection Forms have been detected, are so inaccurate that the defendant should not have relied on these documents. Although both parties agree that it is rather obvious that regarding a number of detected findings of category 1, given the extent to which was deviated from ICAO standards and the effect of these deviations on flight safety, that these findings were in fact only remarks and wrongly considered as findings. However, these category 1 findings only form a subordinate point in the motivation of the contested decision.

The defendant has based the pattern of carelessness and negligence regarding the safety aspects in the contested decision on the number of detected SAFA findings of the second (significant) and third (major) category, and in addition, on a number of incidents and one accident which occurred during the daily operation.

Even after a discussion between experts concerning the detected findings, it is evident that a sufficient number of the detected categories 2 and 3 findings remain persistent to support the contested decision. Shortly prior to the suspension decision of 12 May 2005 it concerned, according to the assessment report of 28 June 2006 of the Dutch Aerospace Laboratory and the supplemental IBA report of 12 July 2007, the following findings:

- During an inspection on 30 April 2005, four category 2 findings were detected on an aircraft of an air carrier called Sky Gate International, domiciled in Kyrgyzstan, and leased by the plaintiff. The plaintiff was obliged to guarantee the safety of this aircraft (a so-called wet-lease-in from a non-JAA operator). One of the findings observed was a Ground Proximity Warning System (GPWS) that did not meet the latest requirements (enhanced GPWS). Moreover, the aircraft showed signs of corrosion.
- Furthermore, there was a crack in the engine compressor inlet section (scrape plate casing) of the right hand engine. IBA stated with respect to the latter finding that this deficiency was only a check-item. This point of view does not convince the court of the fact that the inspectors wrongly identified this finding as a category 2 finding. Finally, a not properly closing waste bin in the lavatory section had to be considered as a category 2 finding.
- During an inspection on 1 May 2005 of an aircraft of Star Air from Sierra Leone, leased by the plaintiff and in that capacity fully responsible for the flight safety requirements, five category 2 findings and one category 3 finding were observed. The findings of category 2 were related to an obsolete type of GPWS (no enhanced type), paint in poor condition as well as visible corrosion, nose gear and left main landing gear shock absorbers leakage, front lavatory leakage and oil leakage in the Auxiliary Power Unit (APU). The category 3 finding was related to the absence of a panel in the engine no 3 cowling section.
- During an inspection on 3 May 2005 (CAA-2005-NL-69) inspectors detected the absence of the registration marks on one side of the aircraft. The remark of IBA that registration marks should be applied with paint instead of stickers does not mean that this finding is mistakenly classified as a category 2 finding.
- During an inspection on 6 May 2005 (CAA-2005-NL-71), a completely worn-out brake unit was discovered, which is a category 3 finding.
- During an inspection on the same date (CAA-2005-NL-72), a category 3 finding was detected consisting of a crack in a panel adjacent to the left hand engine. The IBA has put question marks in classifying the defect as a category 3 finding. However, in the opinion of the court, the IBA has insufficiently substantiated this conclusion.
- During an inspection on 7 May 2005 (CAA-2005-NL-75), three findings of category 2 were identified in an aircraft of the plaintiff.

Of two required available documents no originals or certified copies were at that time present, but only copies of validated copies. The court rejected the argument of the IBA at this point, that the omission did not affect the airworthiness, because the significance of the presence of authentic documents lies in the fact that it can be verified whether compliance with aviation safety standards has been fulfilled.

Furthermore, the fuel in the centre tank, located in the aircraft fuselage, was not included in the calculation of whether or not the maximum allowed zero fuel weight (mass) was exceeded.

- During an inspection on 8 May 2005 (CAA-2005-NL-80), two cockpit crew member smoke-goggles were missing. Furthermore, the waste bin in the front toilet was unserviceable. These findings are both of category 2.
- During another inspection on 8 May 2005 (CAA-2005-NL-82), the incorrect calculation regarding standard passenger weights has been observed, in fact the use of wrong (lower) passenger weights then given the nature of the flight was allowed. At least there was uncertainty as to the question whether the proper procedure was followed. Furthermore, shifting of cargo in two adjacent cargo compartments was detected due to the absence of a barrier net.
- During an inspection on 9 May 2005, the same incorrect calculation procedure regarding standard passenger weights was observed, at least concerning similar uncertainty about the proper procedure to be followed.
- During an inspection on 10 May 2005 (CAA-2005-NL-90), two findings of category 3 were detected, in this case a damaged tyre and erroneously not recording, or at least too late recording, in the maintenance log of a tyre damage.
- During an inspection also on 10 May 2005 (CAA-2005-NL-91), an incorrect serial number of engine no 2 was mentioned in the relevant documents, which is a category 2 finding.

In addition, the defendant may refer, regarding his view with respect to the pattern of negligent and nonchalant handling concerning the safety requirements in the daily operation of the plaintiff's aircraft, to the accident which occurred on 17 June 2003 and the seven incidents that have occurred over a period of two and a half years.[18] Regarding this record, the court takes into account that both parties agree that the occurrence on 8 May 2005 (aborted take-off due to a false fire-warning) should have been classified as a non-culpable incident regarding the plaintiff, and the incident on 2 January 2004 with respect to the loading of an aircraft of the plaintiff at Rotterdam international airport should have been considered as a finding instead of an incident, due to holding incorrect loading documents.

18 Runway overrun after a rejected take-off of an Onur Air McDonnell Douglas MD-88, registration TC-ONP, with flight number OHY 2263 at Groningen international airport Eelde on 17 June 2003 (aircraft seriously damaged, but no casualties).

Regarding the seven incidents, the court is considering the following:

- On 2 May 2003 at Rotterdam international airport an intersection take-off attempt of an aircraft belonging to the plaintiff was prevented by Air Traffic Control.
- The fact that the NLR report should be considered as prejudiced because it has discussed the possible consequences of an intersection take-off, is not evident. After all, the possible consequences are giving an indication concerning the seriousness of the incident.
- On 15 August 2003 at Schiphol international airport, an aircraft of the plaintiff experienced a tyre rupture due to tyre tread separation (category 2 finding), where the inspectors have had to undertake activities in order to move staff members of the plaintiff to properly deal with the damage.
- During a taxi manoeuvre at Eindhoven international airport on 19 August 2003, an aircraft of the plaintiff ended up with the undercarriage off the taxi-track.
- On 26 December 2003 at Rotterdam international airport, an intersection take-off (resulting in dangerously limiting the take-off distance required) attempt initiated by an Onur Air aircraft was again (second time in one year) prevented by ATC.
- On 3 May 2005 at Schiphol international airport, an engine stall incident occurred with an aircraft of the plaintiff. In the opinion of the court the NLR report on this incident has sufficiently substantiated that the crew would depart before all the necessary repairs to the engine were carried out.
- On 5 May 2005, an aircraft of the plaintiff suffering from a seriously damaged tyre departed Schiphol international airport without permission to do so.
- On 10 May 2005, an aircraft of the plaintiff aborted the take-off run at Antalya international airport, located on the south coast of Turkey, on two consecutive occasions, due to a warning signal related to a deactivated engine thrust reverse system, although this warning should have been ignored, according to the dedicated procedure. The opinion of the plaintiff that similar incidents have occurred with other air carriers in the same period does not convince the court. It has not been made plausible that the incidents in those cases originated from one and the same air carrier, or lessons learned from these incidents were inadequate.

As in the Phuket Air judgement, in this particular Onur Air case, the plaintiff argues in vain that the number of findings regarding its aircraft does not deviate from average. The number tends to be even more positive than average. The court judges the opinion of the defendant as being not incorrect to the extent that the defendant was not supposed to be committed to initially assess, in the event that he observes a worrying pattern of negligence and carelessness regarding safety requirements in the daily operation, whether the same or similar worrying patterns exist with other air carriers, prior to taking action. In addition, the court is, contrary to the plaintiff, of the opinion that the defendant has been able to assign decisive weight to the observed pattern of negligence and carelessness, whereby the plaintiff failed to learn lessons from the past, without first having to define in absolute terms what should be understood by an unsafe situation.

In this particular situation, the defendant is entitled to a certain degree of freedom to determine when the pattern of negligence and carelessness will reach a limit beyond which he considers it no longer justified to allow Dutch airspace to be used.

Furthermore, it is essential that a statistic approach, as has afterwards been carried out by the plaintiff, ignores the coherence and context in which the findings are detected and the inspections performed.

Finally, the statistic approach exclusively takes into account the detected findings, including the less important findings of category 1, but this kind of assessment ignores the incidents which took place in a rather short period of time during May 2005. In addition, the defendant should have assigned significance to the fact that the frequency of the holiday charter flights, to be conducted under the operational responsibility of the plaintiff in the currently present summer season, was to be higher than the previous period.

Reviewing all, it is not apparent from the a posteriori formatted expert reports that the defendant has overestimated the seriousness of the problems encountered by the plaintiff regarding safety requirements in general and specifically in the areas of "weight and balance", "wheels, tires and brakes" and reporting and dealing of complaints. In view of the foregoing, the defendant may have come to the conclusion that in the daily operation of the plaintiff a pattern of negligence and carelessness with respect to safety requirements was present in such a way that the continued authorization to use Dutch airspace was no longer justified, and on that basis, the defendant may have proceeded to suspend this route traffic right permit.

Regarding the view of the plaintiff that from findings of the Technical Mission, being deployed to Turkey shortly after the suspension decision (17-19 May 2005), it was clear that the plaintiff's Safety Management System (SMS) was in order, the court however considers that, from the (undated) report of the deployed Technical Mission, it was apparent that the Technical Mission experts were of the opinion that the SMS of the plaintiff was not quite fully in order.

Furthermore, the court determines that the contested decision is not based on an assessment of the plaintiff's SMS in Turkey, but on identified incidents, findings and the aforesaid context in which a few relevant things have played a role.

The plaintiff has rightly taken the view that the defendant should have provided the plaintiff with the opportunity to express reservations prior to the primary decision of 12 May 2005, possibly under the provision of a very short response time. This formal defect has been restored in the objection procedure, because the defendant has taken notice of the plaintiff's concerns during a hearing on 15 June 2005. After that hearing the plaintiff has had the opportunity to express its views in detail, and as such the plaintiff, by not being able to give its opinion, yet is not prejudiced in its interest.

Prior to the suspension of the traffic right permit, no consultation has taken place with the Turkish civil aviation authorities. During the session, the defendant has in an uncontradicted way brought forward that as a matter of fact some consultation with the Turkish civil aviation authorities has taken place.

After taking the primary decision, the Turkish authorities were yet involved in the consultations. The defendant has made it sufficiently plausible that, considering the need for immediate action, a more extensive consultation round with the Turkish civil aviation authorities prior to the taking of the primary decision, in fact was impossible.

Furthermore, the court is considering that the defendant, based on the number of incidents that occurred shortly after each other in May 2005, could reasonably take the view that immediate intervention was necessary on 12 May 2005, yielding sufficient justification for disclosing the motivation of the suspension to the plaintiff within a couple of days, on 17 May 2005.

There is no evidence that the defendant has used its authorization powers for purposes other than to prevent unsafe situations.[19] The fact that the contested decision was actually addressed to the Turkish civil aviation authorities is not adduced convincing evidence.

As far as the appeal is directed against lifting of the suspension of the traffic right permit, it cannot be established what the plaintiff intends to achieve by lodging an appeal, now the defendant, by lifting, is responding to the objections of the plaintiff regarding the decision to suspend, and in pursuance thereof alleged damage. Not in any way is it made plausible that the plaintiff has suffered damage by the lifting decision, or in any other way is affected in its interests. Therefore the plaintiff has no *locus standi* in the appeal case insofar as directed against the enforcement of the lifting of the suspension of the traffic right permit.

Regarding this statement, the court takes into account that the plaintiff has raised no substantive objections against the execution of the provisions included in the contents of the Action Plan.

The plaintiff's position that the Action Plan does not contain provisions that would lead to the improvement of the interaction with safety requirements in the daily operation of the plaintiff – whatever significance it even might give – does not detract from the fact that the plaintiff is not adversely affected by the lifting of the suspension. The court dismisses the appeal, insofar as it is directed against the enforcement of the decision to lift the suspension.

The appeal, insofar as it concerns the enforcement of the decision to suspend, is admitted for misapplication of Article 3 of the BOL. The contested decision insofar must be annulled. However, the court shall maintain the legal effects of the annulled decision, because the court deemed the defendant competent, on grounds previously been determined, to suspend the plaintiff's traffic right permit.

19 Détournement de pouvoir. Abuse of power.

Furthermore, the court considers that the suspension does not substantially infringe with one or more written or unwritten principles of good governance or general principles of justice.

9.5 Decision of the Court

Hereby the court:
- dismisses the appeal, insofar as it is directed against the enforcement on the lifting of the suspension of the traffic right permit.
- declares the appeal well-founded, insofar as it is directed against the enforcement of the suspension of the traffic right permit.
- shall annul the contested decision of 27 July 2005, identification number JZ/4148.0505 LVW, insofar as
- determines that the legal effects of the annulled part of the decision will remain intact.
- (...).

9.6 Concluding Remarks

Safety information management, in the broadest sense of the term, the collection and analyzing, the exchange, protection and dissemination of informative data, is widely used in various organizational structures, but above all in civil aviation, as a communication system to prevent recurrence of accidents by identifying potential deficiencies in the entire organization, human as well as technical, at an early stage. A proactive approach in the ever advancing technological development is an indispensable factor, yet it requires a complex adjustment and effort to cope with difficult controllable operating activities.

Civil aviation authorities and air carriers audit data, as well as aircraft assessment reports of SAFA ramp inspections, stored in the centralized EASA database, all constitute key elements and are playing a crucial role in ensuring a high level of safety in civil aviation.

To protect air travellers around the world, in general, but regarding the SAFA intent, more particularly to warn European air travellers, but also regarding foreign air carriers to ward off potential disasters, the European Union introduced the real option for imposing an operating ban within the Community on those air carriers that are not complying with international aviation safety requirements.

However, not without striking a blow, witness the fact of several legal disputes between EU member State governments and foreign operators regarding bans imposed by those individual governments in the Community.

A typical example of the discontent of an air carrier with the measures regarding an operating ban, imposed in 2005 by a government, was the legal dispute between Phuket Airlines and the Dutch authorities. Phuket Airlines disagreed with its traffic permit revocation by the Dutch Government, which was based on a number of SAFA inspection reports revealing most serious findings and subsequent failure to ensure corrective actions. There was a significant difference of opinion about revocation or suspension of air traffic permits. Regulatory law in this dispute consisted of the Chicago Convention (especially Article 11), the Air Traffic Agreement (1971) between the Government of the Kingdom of Thailand and the Government of the Kingdom of the Netherlands, and the Dutch Aviation Act.

A similar legal dispute concerning suspension of traffic rights was the case with Turkish air carrier Onur Air on operating holiday charter flights between the Netherlands and Turkey. The air carrier operates scheduled domestic air services and holiday charter flights to a large number of destinations throughout Europe. Poorly performing, to some extent rather unsafe, daily operations of this air carrier, determined by subsequent SAFA inspections in 2005 and as early as 2003 when an Onur McDonnell Douglas MD-88 overshot the runway of Groningen international airport Eelde, led the Dutch authorities to decide, in view of safety, to suspend the traffic rights of Onur Air to Dutch destinations for a month so as to set things in order.[20]

Political pressure from the Turkish Government remarkably forced the Dutch authorities to reduce the duration of the suspension case. Attempts by Onur Air to receive any compensation because of alleged losses after its banishment, were overruled by the Dutch court in appeal.

20 Switzerland, France and Germany also denied Onur Air access to their respective airspaces for alleged safety breaches.

10 Safe Transport of Dangerous Goods by Air

10.1 Legal Aspects

In the SAFA aircraft inspection cycle, on-board dangerous goods are a significant safety check item, especially the loading, the nature, the labelling and packaging method of those hazardous materials.

More than half of the cargo carried by all modes of transport in the world is considered to be dangerous cargo. These dangerous goods, such as explosive, corrosive, flammable, toxic and even radioactive material, are essential for a wide variety of global industrial, commercial, medical and laboratory research activities, requirements and processes. Because of the great advantages of air transport, specialized and fast (inter)continental on-time delivery, using only a single means of transport over long distances, a great deal of this dangerous cargo is carried by aircraft.

To ensure that dangerous goods cargo can be carried safely by air, ICAO adopted Annex 18 in 1983. This Annex is somewhat unique in that it reflects coordinated efforts with the United Nations and the International Atomic Energy Agency (IAEA), ensuring full compatibility with their Technical Instructions. ICAO Annex 18 specifies the broad SARPs to be followed to enable dangerous goods to be carried safely by air. The Technical Instructions contain the very detailed and numerous instructions necessary for the correct handling of dangerous cargo.[1]

A comprehensive set of requirements provide for the classification and a complete list of dangerous goods. The list identifies those goods that are:

a. Forbidden under any circumstances;
b. Forbidden on both passenger and cargo aircraft in normal circumstances but could be carried in exceptional circumstances subject to the exemption by the States concerned;
c. Forbidden on passenger aircraft but permitted on cargo aircraft in normal circumstances; and
d. Permitted on both passenger and cargo aircraft in normal circumstances.

The ICAO requirements for dangerous goods have been developed largely by a panel of experts established in 1976.[2] This panel is a major authority in continuing to meet and recommend the necessary revisions to the Technical Instructions, which in turn will be

1 ICAO Doc 9284 Technical Instructions for the Safe Transport of Dangerous Goods by Air.
2 ICAO Annex 18 came into force on 1 January 1984, much later than some of the original Annexes. Annex 6, for example, dates from 15 July 1949.

kept in line with the Recommendations of the United Nations Committee of Experts on the Transport of Dangerous Goods and with the Regulations for the Safe Transport of Radioactive Materials of the IAEA.[3]

The successful application of regulations concerning dangerous goods in air transport greatly depends on the appreciation by all individuals concerned of the risks involved and on a highly detailed understanding of the relevant regulations, extremely important with respect to a possible infringement of safety. While all Annexes to the Chicago Convention are normally directed to contracting States, Chapters 7, 8 and 9 of Annex 18 include obligations for the shipper and the operator with references to the Technical Instructions.

Essential principles are adopted in order to carry dangerous goods safely on board an aircraft. These principles have been used in developing the Technical Instructions. They are intended to facilitate transport while providing a level of safety such that dangerous goods, providing all the requirements are fulfilled, will under normal conditions not give rise to uncontrollable situations.

The Technical Instructions are designed to ensure, should an incident occur, that the arisen situation cannot lead to an aggravating case like a serious accident.

Annex 18 is an important document with a wide range of implications and certain critical aviation safety matters, applicable not only in the air but also on the ground. An aircraft accident, involving dangerous goods, of whatever nature, may have serious and far-reaching consequences not only for aircraft occupants but also for people on the ground as well as for the environment.

Expertise on the characteristics and consequences of dangerous goods is of utmost importance, not only at major air carrier levels but equally at regional levels in developing States. Governments are supposed to monitor compliance levels at both air carrier and aerodrome levels. Aerodrome management certainly has a big role to play as custodian of dangerous goods on their premises.

Storage of different categories of dangerous goods together and in the company of other goods may normally not be dangerous, if in compliance with the Technical Instructions and regulations from either ICAO or IATA, but catalyzed by certain other dangerous goods may form the foundation for aerodrome structural fires and other environmental calamities, let alone the disastrous impact in case of a crash of an incorrect, highly dangerous loaded aircraft in a populated area. With respect to passenger aircraft, there are obviously more restrictions regarding the acceptance of dangerous goods for transport.

Dangerous goods shall not be carried in an aircraft cabin occupied by passengers or on the flight deck of an aircraft, except in circumstances permitted by the provisions of the Technical Instructions, for example on board Combi aircraft under specific rules and limitations.[4]

3 ICAO Annex 18, 3rd edn, July 2001, The Safe Transport of Dangerous Goods by Air, Foreword.

4 *See also* ICAO Annex 18 8.5 Loading restrictions in passenger cabin or on flight deck.

10.2 Hazardous Materials as a Cause of Aviation Accidents

A catastrophic accident, most probably caused by an in-flight fire in the cargo area on the main deck, was South African Airways Flight 295. On 28 November 1987, SAA Flight 295, a Boeing 747-244B Combi aircraft, crashed into the Indian Ocean about 200 NM east of the island of Mauritius, killing all 140 passengers and 19 crew members, on a commercial flight from Taipei Chiang Kai-Shek international airport in Taiwan, Republic of China (ROC), to Jan Smuts international airport (now OR Tambo international airport), Johannesburg, in the Republic of South Africa via the stopover place Plaisance international airport (now Sir Seewoosagur Ramgoolam international airport) at Mauritius.[5] Nine hours after departure, with only an hour remaining flight time before reaching Mauritius, the smoke alarms went off, signalling smoke in the cargo hold at the main deck, directly behind the passengers compartment. The cargo fire, originating from the front right-hand cargo pallet, spread rapidly. Whilst the crew tried to extinguish the fire, passengers were having trouble breathing through the toxic fumes. An emergency descent was initiated by the crew, but remedial actions caused adverse effects. As the crew struggled to land the plane safely, some passengers died from smoke inhalation. Three minutes after the last contact with ATC, the Boeing crashed into the Indian Ocean, killing those passengers and crew who were still alive. This catastrophic cargo fire was one of the few that have occurred on wide-body aircraft and the first ever on a 747 Combi. After the tragic accident SAA discontinued its use of the 747 Combi, and the FAA (the aviation authority in the State of Design and the State of the Manufacturer) created in 1993 new standards for the 747 Combi, which made it economically non-viable.

The reason for the loss of Flight 295 was not specified. The aircraft accident investigation report detailed two possibilities:

1. The cockpit crew became incapacitated by smoke penetration into the cockpit;
2. The fire weakened the aircraft's fuselage structure, and the tail section separation very likely caused the aircraft to be uncontrollable.

The cause of the fire could not be determined. The exact source of ignition was never identified, but the report concluded that sufficient evidence was found to confirm that the fire had burned for some time, even fuelled by oxygen, and that it might have caused structural damage. An official investigation commission (Margo commission, South Africa) concluded that it was impossible to allocate blame to anyone for the fire, thereby abstaining from rumours of possible terrorist acts or explosive devices, or weapons carried on board.

Despite forensic examinations and allegedly conspiracy theories, no new evidence was found to justify reopening of the inquiry. However, irregularities in the CVR transcript, missing ATC tapes, an arms embargo in the Republic of South Africa at that time,

5 The Boeing 747-244BA Combi (short for combination) aircraft is a class of aircraft that allows for cargo and passengers to be loaded on the same deck (There are cargo compartments below the passenger deck as well as behind the last row of passengers on the main deck).

concealed clandestine weapon transport, cover-up actions, biased participation of the South African CAA in the official investigation are, however, unproven but challenged aspects of this bizarre cargo (dangerous goods) fire accident.

Therefore, ICAO's Annex 18 is one of those Annexes whereby not "playing by the rules" of a State in one part of the globe may have very serious consequences on aviation safety in a totally different part of the world. In this respect its impact is similar to that of Annex 6 or 8. It is reiterated that wrongly managed dangerous goods in one State may be transported by air to different States, transshipped or temporarily stored at various airports. If anyone in the overall transportation process from the original packer to the final consignee, in fact the weakest link, does not follow the provisions of Annex 18 and those of the Technical Instructions, it could potentially endanger not only the flight segment, with possible disastrous consequences, but also transit points. Moreover, manufacturers and regulators must do their utmost to especially prevent airborne cargo fires. Installation of automatic fire extinguishing systems in the hold of cargo aircraft, smoke hoods, strict regulations and instructions concerning prevention, detection and suppression of cargo fires are most important, not only for the cargo flight crew to survive safely but moreover, and to a great extent, to protect vulnerable populated areas and the environment.[6]

Operators are aware of what kind of dangerous goods have been loaded on their aircraft. In case of an aircraft accident the Technical Instructions require that they must, as soon as practicable, inform the State in which territory the accident occurred, about the nature and quantity of these goods and their location on board.

However, it is quite possible that, depending on the circumstances and location of an accident, this information may not be available instantly. Moreover, some accident investigation reports revealed alleged fraud or cover-up (SAA flight 295), although in most cases this was reduced to confusion or errors owing to lack of clarity about cargo data.[7]

---

6 United Parcel Service Airlines flight 6, a Boeing 747-400F, flying between Dubai international airport and Cologne/Bonn international airport, crashed on 3 September 2010 near Dubai shortly after take-off, killing the two crew members. A cargo fire was determined as the probable cause. UPS flight 6 contained a large quantity of lithium batteries. According to studies of the NTSB, the agent type Halon 1301 was inefficient in fighting fires involving these specific dangerous goods.

7 "Een Beladen Vlucht", Final Report of the Dutch Parliamentary Enquête (Survey) Commission on the Bijlmer-crash of 22 April 1999 (El-Al flight LY 1862 on 4 October 1992 near Amsterdam, the Netherlands). *See also* the confusion regarding the Lufthansa flight LH 8460 accident (in-flight – false – cargo compartment fire warning, declared emergency (later denied), and crash landing) at King Khalid international airport, Riyadh, in Saudi Arabia on 27 July 2010. The German news magazine *Der Spiegel* reported on 14 August 2010 that ground witnesses at the departure location, Frankfurt international airport, confirmed the cargo contained highly inflammable chemicals as well as other hazardous goods like weaponry and other military goods. According to that report the chemicals were located exactly in the area where the fire started. Lufthansa Cargo did not comment on that report. German investigators (State of Registry) of the German Federal Bureau of Aircraft Accidents Investigation (*Bundesstelle für Flugunfalluntersuchung, BFU*) disclosed that the Lufthansa Boeing (Mc Donnell Douglas) MD-11 freighter which crashed and burned on landing in Riyadh made three heavy contacts (pilot-induced) with the runway (*source:* Flight International, 17 September 2010). *See also* BFU 2X003-10.

According to the U.S. NTSB (State of the Manufacture), participating in the accident investigation, led by the General Authority of Civil Aviation Saudi Arabia (GACA), an uneventful cruise flight and approach excessive sink rate during the landing procedure and subsequent bouncing on the runway caused the fuselage to rupture and catch fire (NTSB Safety Recommendation, 12 July 2011).

According to the requirements of the Technical Instructions, operators must report to the relevant authorities any accident or incident involving dangerous goods. States in turn are obliged to have procedures in place to investigate such occurrences.

Another tragic accident caused by hazardous materials on board a passenger aircraft was ValuJet flight 592. A McDonnell Douglas DC-9-32 aircraft with registration N904VJ, operated by ValuJet Airlines, an American air carrier, a low-cost service (no-frills basis, low fares) company based in Atlanta, was dispatched for a domestic passenger flight, under the provisions of 14 CFR (Code of Federal Regulations) Part 121, between Miami international airport, Florida and William B. Hartsfield Atlanta international airport, Georgia.

The aircraft crashed approximately 10 minutes after take-off into the Everglades, just 17 nautical miles northwest of Miami, Florida, USA, on Saturday 11 May 1996.

During the climb out of Miami a flight attendant, alarmed by fire and smoke in the passengers cabin, but unable to contact the flight crew by intercom, entered the cockpit to inform the pilots about the untenable situation. The cockpit already suffered from unidentified electrical problems, and action was taken immediately. The aircraft was cleared for an immediate return to Miami, but a few minutes later Flight 592 disappeared from the radar and crashed in the Browns Farm Wildlife Management area in the Everglades. Both pilots, the three flight attendants and all 105 passengers were killed instantly.

The National Transport Safety Board (NTSB) determined that the probable (underlying) causes of the accident, which resulted from a fire in the airplane's class D cargo compartment, a compartment in which fire suppression is accomplished by sealing off the hold from outside air,[8] in fact an uncontained fire that was initiated by the actuation of one or more chemical oxygen generators being improperly carried as cargo,[9] were:

1. The failure of SabreTech Inc. (aircraft maintenance contractor) to properly prepare, package, and identify unexpended chemical oxygen generators before presenting them to ValuJet for carriage;

---

8 A Class D cargo hold, according to Federal Regulation 14 CFR 25.857 (d), is a compartment in which a fire will be completely confined without endangering the safety of the airplane or occupants, the smoke or hazardous gases from any fire therein are confined to the cargo hold, the fire is likely to die from oxygen starvation and the compartment is <1,000 cubic feet.

9 Oxygen generator canisters are commonly installed in passenger aircraft to provide a source of oxygen to occupants in case of a sudden loss of cabin pressure. The canisters used in McDonnell Douglas MD-80 aircraft series contained spring-loaded igniters secured by a safety device (retaining pin). In an emergency a passenger will pull down on the oxygen mask normally located in the overhead compartment, which is protected by a heat shield, and as a result the pin is pulled out of the canister. With the pin removed, the spring-loaded igniter will strike the canister which in turn creates a chemical reaction that produces oxygen. The amount of heat produced by the generators during this process, especially in unshielded conditions, makes them extremely dangerous and considered to be hazardous materials.

2. The failure of ValuJet to properly oversee its contract maintenance program to ensure compliance with maintenance, maintenance training, and hazardous materials requirements and practices; and
3. The failure of the FAA to require smoke detection and fire suppression systems in class D cargo compartments.

Contributing to the accident was the failure of the FAA to adequately monitor ValuJet's heavy maintenance programmes and responsibilities, including ValuJet's oversight of its contractors, and SabreTech's repair station certificate. The NTSB, in its final accident investigation report, considered more specific shortcomings on the side of the U.S. Government and the company: the failure of the FAA to adequately respond to prior chemical oxygen generator fires with programmes to address the potential hazards, and ValuJet's failure to ensure that both ValuJet and contract maintenance facility employees were aware of the carrier's "no-carry" hazardous materials policy as well as its failure to effectuate that these employees received appropriate hazardous material training.[10] In other words, more generally the FAA was lax in monitoring and in enforcement of air cargo safety regulations while ValuJet, as a start-up "lean and mean" air carrier, as a direct product of governmental deregulation policy, was negligent with monitoring outsourced maintenance and other critical support functions and seemed to have developed a culture that conflicted with flight safety.

In case studies, the crash has been described as the result from the "mutually reinforcing interaction" between private corporations, *in casu* ValuJet and SabreTech and a governmental agency, the FAA, and as such representing a model of State-facilitated corporate crime in which pursuit of profit, disregarding of safety measures and the failure of a State agency to effectively monitor air carrier adherence to maintenance and dangerous goods regulations, which resulted in a, however avoidable, horrible deadly crash.

The oxygen generators, originating from three purchased McDonnell Douglas MD-82/83 aircraft, had to be inspected by SabreTech's maintenance and overhaul facility in Miami as part of an ongoing contractual relationship with ValuJet for line maintenance support and heavy aircraft maintenance.[11] The inspection of the oxygen generators on request of ValuJet was necessary to determine whether they had exceeded the allowable service lifetime of 12 years from the date of manufacture, and if expiration was a fact, to replace the generators concerned.

Some expired generators had to be shipped to Atlanta for disposal. The correct procedure before disposal (and shipping) is to discharge the generators, and even discharged

---

10 NTSB Aircraft Accident Report PB97-910406 NTSB/AAR-97/06 DCA96MA054, *In-Flight Fire and Impact with Terrain ValuJet Airlines flight 592 DC-9-32, N904VJ, Everglades, Florida, May 11, 1996*, adopted 19 August 1997 notation 6704B.

11 The accident plane (DC-9) did not have a chemically generated oxygen system installed in its cabin. None of the DC-9-30 series airplanes use chemically generated oxygen systems.

generators are identified as hazardous waste and require as such associated treatment as labelling and packaging (caps installed), as required by law.

Because of miscommunication, as well as negligent acting and careless handling regarding a number of operational but unexpended oxygen generators, without proper discharging, packaging, labelling and a meticulous inspection, these generators and non-hazardous empty units (canisters) were eventually packed together in boxes, and after sealing marked as COMAT (Company Materials) and listed as Oxygen-Canisters Empty, to be loaded on board the DC-9. On the basis of information available, the flight crew would have had no reason to know or suspect that hazardous materials were being proffered for carriage on board the airplane.

The COMAT, the boxes with the assumed non-hazardous canisters, together with some tyres, was not secured inside the cargo compartment by netting straps, or any other means to prevent movement of the items because of turbulence or excessive aircraft manoeuvring. The hazardous materials regulations require that packages containing hazardous materials must be secured. Although it is possible that the generator(s) that initiated the fire actuated as a result of being struck by unsecured cargo (such as tyres), and that the accident could have been avoided by securing the cargo, it cannot be conclusively determined whether unsecured cargo did play a role in actuating the uncapped generators.[12]

Some of the relevant conclusions of the NTSB in its final report with regard to the handling and transportation of dangerous goods in the ValuJet case are listed below:

1. The accident airplane was equipped and maintained in accordance with Federal Regulations and approved procedures, and there was no evidence of pre-existing mechanical malfunctions or other discrepancies in the airplane structure, flight control systems, or power-plants that would have contributed to the accident.
2. Weather was not a factor in the accident.
3. The activation of one or more chemical oxygen generators in the forward cargo compartment of the airplane initiated the fire on ValuJet flight 592. One or more of the oxygen generators were likely actuated at some point after the loading process began, but possibly as late as during the airplane's take-off roll.
4. Even if the fire did not start until the airplane took off, a smoke/fire warning device would have more quickly alerted the pilots to the fire and would have allowed them more time to land the airplane.
5. If the plane had been equipped with a fire suppression system, it might have suppressed the spread of the fire (although the intensity of the fire might have been so great that a suppression system might not have been sufficient to fully extinguish the fire) and it would have delayed the spread of the fire, and in conjunction with an early warning, it would likely have provided time to land the airplane safely.

12 NTSB Aircraft Accident Report (note 350), at 105.

6. Had the FAA required fire/smoke detection and fire extinguishing systems in class D cargo compartments, as the Safety Board (NTSB) recommended in 1988, ValuJet flight 592 would likely not have crashed.
7. The loss of control was most likely the result of flight control failure from the extreme heat and structural collapse; however, the Safety Board cannot rule out the possibility that the flight crew was incapacitated by smoke or heat in the cockpit during the last 7 seconds of the flight.
8. Given the potential hazard of transporting oxygen generators, and because oxygen generators that have exceeded their service life are not reusable, they should be actuated before they are transported.
9. Had a warning label or emblem clearly indicating the significant danger posed been affixed to each generator, personnel handling the generators, including the personnel in shipping and stores who prepared them for shipment to Atlanta, might have been alerted to the need to determine how to safely handle and ship the generators.
10. Improper maintenance activities and false entries pose a serious threat to aviation safety and must be curtailed.
11. The failure of SabreTech to properly prepare, package and identify the unexpended oxygen generators before presenting them to ValuJet for carriage on board Flight 592 was causal to the accident.
12. The procedures of many air carriers for handling hazardous company materials (COMAT) are not fully consistent with the hazardous materials regulations and the guidance provided on 13 December 1996 by the Research and Special Programs Administration on the transport of COMAT by air carriers.
13. ValuJet failed to adequately oversee SabreTech and this failure was a cause of the accident.
14. Before the accident, the FAA's oversight of ValuJet did not include any significant oversight of its heavy aircraft maintenance functions. The FAA's inadequate oversight of ValuJet's maintenance functions, including its failure to address ValuJet's limited oversight capabilities, contributed to this accident.

In summary, the NTSB found that the failure to install the shipping caps was a critical error, because if they had been properly installed, the oxygen generators would not have been actuated and the accident would not have occurred.[13]

Apart from the NTSB investigation and Congressional hearings, SabreTech and its employees were accused of criminal liability for the crash. The U.S. Federal and Florida State Prosecutors brought criminal charges, multiple manslaughter and third degree murder against ValuJet maintenance contractor, SabreTech, its maintenance supervisors and two mechanics who worked on the aircraft.

---

13 J.M. Walters & R.L. Sumwalt III, *Aircraft Accident Analysis: Final Reports*, McGraw-Hill Professional, New York, NY 2000, at 116.

Furthermore, SabreTech was charged with violation of hazardous materials regulations, failing to train its personnel on proper and safe handling of hazardous materials, placing a destructive device on board a commercial (passenger) aircraft, conspiracy to falsify records and making false statements.[14]

Eventually, in 2001, a Federal Appeal Court overturned the convictions and reversed the fines. State murder charges were dropped in exchange for a donation to promote aviation safety. SabreTech and ValuJet dealt with victim lawsuits as well.[15]

In the context of SAFA inspections in the European Union, aircraft cargo compartments and dangerous goods are subject to inspection, largely because of lessons learned in the past. Despite the limited nature of SAFA ramp inspections, it is of utmost importance, in view of the flight safety critical characteristics of hazardous materials, to have a (rather time consuming) comprehensive check on labelling, packaging and stowing, NOTOC (Notification to Captain) and cargo manifests concerning dangerous goods. In case of uncertainty or suspicion it would certainly trigger the attention of SAFA inspectors to thoroughly check the cargo compartments and selective cargo (dangerous goods).

10.3 CONCLUDING REMARKS

Civil air transport is by far the most expeditious way of moving passengers and cargo, generally time-dependent goods, around the world from one distant point to another. Worldwide economic growth has contributed to massive transport of any goods by air, including hazardous materials and substances. Presently, dangerous goods are regularly and routinely carried as cargo on board passenger and all cargo aircraft. These goods pose little risk to air transport provided they are carried in limited quantities, correctly identified, packaged and handled.[16]

Experience from other forms of transportation, in conjunction with the knowledge gained by the aviation industry of the characteristics of air transport, resulted in the first industrial regulations on the safe transportation of dangerous goods. IATA, as a global air transport organization, has drawn up the IATA Dangerous Goods Regulations (IATA DGR).

These regulations are, however, based on the requirements of Annex 18 to the Chicago Convention and the subsequent editions of ICAO Technical Instructions (ICAO TI), which are the legal documents relating to air transport of hazardous materials (dangerous goods).

---

14 S. Michaelides-Mateou & A. Mateou, *Flying in the Face of Criminalization. The Safety Implications of Prosecuting Aviation Professionals for Accidents*, Ashgate Publishing Limited, Farnham 2010, at 171.

15 R.W. Cobb & D.M. Primo, *The Plane Truth, Airline Crashes, the Media, and Transportation Policy*, Brookings Institution, Washington DC 2003, at 98.

16 ICAO defines dangerous goods as articles or substances that are capable of posing a risk to health, safety, property or the environment and that are shown in the list of dangerous goods in the Technical Instructions or that are classified according to those Instructions.

As in all potentially hazardous but generally strict regulated activities, such as commercial air transport, awareness is the key factor in maintaining a high level of safety. In 1996, the world sadly saw the tragic consequences of insufficient awareness and negligence when a ValuJet DC-9 fatally crashed into the Florida Everglades after hazardous materials had caused a fierce, irrepressible fire on board.

It is essential for the public, especially air travellers and people on the ground exposed to potential aviation-related risks, but as well to governments and the aviation industry, to continually work towards minimizing of risks involved in the transportation by air of these hazardous materials and substances through the application of safe practices, such as (SAFA) inspections based on appropriate regulations.

11 Aviation Safety in the Vicinity of Community Airports

11.1 Introduction

Within the framework of the Community's overall safety-improvement strategy to establish and maintain a high uniform level of civil aviation safety in Europe, Directive 2004/36/EC on the safety of third-country aircraft using Community airports introduced a harmonized approach to effectively enforce international safety standards by the member States within the Community by harmonizing the rules and procedures for ramp inspections of third-country aircraft landing at airports open to international air traffic and located in the member States.[1]

To protect the interest of the European citizens who may be living in the vicinity of Community airports or will be travelling on board of third-country aircraft, the Community identified the need to effectively enforce international safety standards within itself. It must be assumed that Community air carriers and their aircraft fully comply with these international aviation safety standards, or even more stringent requirements dictated by civil aviation authorities of the EU member States, and thus to ensure an as-safe-as-practicable flight performance.

> The exposure of EU citizens on the ground to the consequences of aircraft accidents is particularly unacceptable. The recent increases in the number of people killed on the ground by aircraft in the vicinity of airports have increased public awareness of this risk. Third-Party Risk (TPR) around airports is becoming a major limitation to airport-expansion plans intended to keep pace with increased demands for capacity. Therefore, decreasing the likelihood of accidents is a high-priority issue for airport regions, and measures to improve should take into account not only aircraft passengers and crew but also third parties.[2]

Aircraft, not particularly complying with these international safety standards, will nevertheless land at airports located in the member States. It means that aircraft with severe

1 Article 1 Scope and objective of Directive 2004/36/EC.

2 SEC (96) 1083-C4-0423/96, 20 March 1997. Report on the Commission's communication on defining a Community aviation safety improvement strategy.

deficiencies, which in some cases could lead to dangerous situations, could, to a much greater extent than Community aircraft, be potentially hazardous to citizens and property in the vicinity of airports.

One of the tools to constrain these ubiquitous hazards is the execution of ramp inspections on third-country aircraft operating to and from Community airports. Inspections, revealing potential aviation-safety problems, eventually will ensure compliance with safety requirements and strict safety oversight in all respects, thus mitigating the risk of fatal accidents in the future. The aviation industry benefits from a high level of safety realized not only by air carriers but also by all parties involved in airport operations. A fatal accident at or near an airport could not only harm the safety reputation of the air carrier in particular, and air transportation in general, but also be detrimental to the image of the airport involved and above all shocking to (vulnerable) people living around airports. At least their confidence in aviation safety can be seriously affected.

Resolution B4-0150/96 on the air disaster off the coast of the Dominican Republic, adopted by the European Parliament on 15 February 1996 (OJ C 65, 4.3.1996, at 172) highlights the need for the Community to take a more active stance to submit practical proposals for increasing the safety of civil aviation. In particular, in this document, the Parliament is calling on the European Commission to develop a strategy to improve the safety of its citizens travelling by air or living near airports.[3]

Community Regulations such as Regulation (EC) No 2111/2005 and Directives, like Directive 2004/36/EC, regarding aviation safety, have been established in the European Union to enhance aviation safety, to protect public safety, citizens and property on the ground and passengers on board of commercial aircraft, being both foreign aircraft as EU member State aircraft.

Many major European international airports are located in extremely densely populated suburban areas despite strict noise-abatement rules, which purport the least inconvenient and most gentle low-risk approach and departure procedures. As a consequence, air traffic routes to and from these airports inevitably concentrate over these areas, and, as statistics plainly indicate, aviation accidents mainly occur during the take-off, approach and landing phase of flights.

Around 25% of all aircraft accidents occur during the take-off phase and approximately 50% during the approach and landing phase, thereby demonstrating that air transport safety is closely related to airports and their immediate surroundings. Aviation activities around major airports are considered high-risk activities, however regulated to minimise the risk of accidents or serious incidents.

---

3 Directive 2004/36/EC of the European Parliament and of the Council of 21 April 2004 on the safety of third-country aircraft using Community airports, under (1).

The likelihood that an aviation accident or serious incident occurs is relatively small. However, if an aircraft accident happens, it might be disastrous for the aircraft occupants, but at the same time the consequences for residents and property on the ground literally and figuratively could have the effect of a major impact.

Since the Bijlmer-disaster on 4 October 1992, a crash of an El-Al Boeing 747 freighter in a densely populated Amsterdam suburb, the awareness regarding safety in the vicinity of major airports has increased significantly. From that time on, the concept "external safety" has been introduced in the Netherlands, but, in fact, will apply everywhere, and will be part of the decision making with regard to the organizational installation layout and operational requirements of major airports. In this respect, the Dutch Aviation Act allows to incorporate a limit in the Airport Resolution of a major airport.

External safety policy has the objective of protecting vulnerable locations such as housing, schools, hospitals, businesses, offices and so on in the vicinity of airports.

External safety focuses on the effects directly caused by an accident of an aircraft in the take-off or approach-to-land phase, in which persons outside the airport perimeters are exposed. External safety especially concerns the risk of local residents in the vicinity of an airport rather than people who are present at the airport (internal safety). External safety focuses on managing these risks.

External safety not only applies to land use and infrastructural planning systems around busy airports but also applies significantly to the regulation of air transport in such a way to keep the total risk within tolerable limits, as complete elimination of total risk in modern society is virtually impossible. This chapter describes the regulation concerning the safety around major airports (external safety of aviation) and the preparedness for possible aviation-related accidents at or around airports in the Netherlands.

11.2 Probability of an Aviation Accident at a Particular Location

The Dutch policy with respect to external safety of aviation is based on models that give a prediction of the likelihood of an aviation accident at a particular location. Three elements in the external safety model are as follows:

- Accident probability (determining the risk of an aviation accident per accident type based on historical statistic data),
- Accident site (determining the probability that an aviation accident occurs at a particular location. Locations of aviation accidents are assessed from historical data) and
- Accident consequences (determining the consequences of an aviation accident at a particular location based on historical data).

The consequences of an aviation accident are not confined to the crash location, but are felt in a restricted zone around that location as well. The accident consequences are modelled by two parameters: the extent of the area exposed to the consequences of an aviation accident and the lethality. The dimensions of this area depend on the type of terrain as well as the size of the aircraft involved (expressed in MTOM).

An additional factor may be the cause of the accident (in-flight explosion or disintegration) and the subsequent mode of impact (compact or widely scattered).

Because of the multitude of assumptions that necessarily are to be present in the model, this specific model is suitable only for predicting the likelihood of aircraft accidents in diverging areas covering the extended centrelines of take-off and landing runways, but is not suitable for taking measures in the area planning sphere in a region above which (low-altitude) airway or high-level jet routes are situated.

11.3 DISTINCTION BETWEEN LOCATION-RELATED RISK AND SOCIETAL RISK[4]

People on the ground near the ends or extended areas of airport runways, living there, working there or passing through, all for reasons unrelated to aviation, are subject to an increased risk of death or serious injuries caused by (civil) aviation activities, in the worst-case scenario by possible aircraft accidents. This risk is called Third-Party Risk (TPR) or external safety risk.[5] Regarding the external risk model, there are two distinct types of risks: the individual (location-related) risk and the societal (group) risk. DEGAS emphasizes in its exploratory study for advice 2011-034 "*Veiligheid als deel van het geheel, een rationele bedding voor emotie*" (Safety as part of the whole, a rational course for emotion) to the Dutch Parliament, that location-related risk will tell us about the safety of people being present around an airport, and that societal risk will tell us something about the safety of society as a whole. External safety risk can be distinguished from internal safety risk, which is entirely related to aviation activities.

Internal safety in the aviation industry is at a high uniform level and still gradually improving through (inter)national rules and regulations and enforcing compliance (SAFA, EU blacklist, IASA, USOAP, etc.) with internationally agreed safety requirements, and at governmental-level State safety programs as well as at operating level safety management systems, in which traditional reactive processes will be complemented by proactive components, whereas improvement in external safety has to come from strict measures taken by governments in the field of land use and infrastructural planning, apart from habitat protection commitments, in case of balancing of interests and reducing activities possibly interfering with or inferior to civil air transport.

---

4 *See also*: Dutch Expert Group Aviation Safety (DEGAS) Advisory report 2011-034, presented on 4 July 2011, at 3.

5 Third Party Risk is the chance that people will die from the impact of an external object (a crashing aircraft), without being part of it. *See also* DEGAS Advisory Report 2011-034, paragraph 2 The concepts, at 9.

Individual (Location-Related) Risk

The individual risk, which nowadays is called location-related risk (the risk linked up with a certain location related to the airport's position and its air routes system), where aviation is considered as one of various activities that influence the pretended location safety, describes the probability per year that an imaginary person, who for a period of one year permanently resides in the same place, dies as a result of an aviation accident.[6]

The risk profile increases as the distance to the departure or arrival routes (as well as to designated runways at the airport itself) decreases.

Connecting points on the ground with the same risk produce a specific contour (the so-called location-related risk-contour) not unlike noise level contours. An average probability of once in 100,000 years would be recorded as 10-5.

Societal (Group) Risk

Location-related risk does not tell anything about how many people are residing or present in a high-risk area, and does not tell anything about social disruption caused by an aviation accident. However, it is the concept of societal risk, which will give an indication about this issue.

Societal risk concerns the probability per year that in one instance a group of at least a certain size on the ground will die from the effect of an aviation accident. Such a tragic event will have an enormous impact on society. Other concepts regarding societal risk are based on the fact that this kind of risk concerns societal effects and harming or degradation of society in general.[7] In determining the societal risk, the presence of groups of people in the vicinity of an airport will play an important role. However, the level of societal risk calculations (the calculations of TPR around airports) may differ depending on the methods, models and implementations used in specific cases.[8]

The societal risk will indicate, given the number of people residing in the vicinity, the probability that a group of persons will die due to an accident in a hazardous activity. According to the Regulation External Safety Facilities (*Besluit Externe Veiligheid Inrichtingen BEVI*), the societal risk is the cumulative annual probability that at least 10, 100 or 1,000 persons will die as a direct result of their presence in an area affected by a facility and an unusual incident within this facility involving a hazardous substance (dangerous goods).

6 Individual risk is the annual risk of death or serious injury to which specific individuals are exposed (R. Wood, Airport Watch, November 2010).

7 DEGAS Advisory Report 2011-034, at 9.

8 M.A. Piers, 'Models for the Assessment of Third Party Risk due to Aircraft Accidents in the Vicinity of Airports', *in* A. Mosleh & R.A. Bari (Eds.), *Probabilistic Safety Assessment and Management*, Springer-Verlag London Ltd., London 1998, at 1552.

Another risk-assessment methodology regarding external safety is the Potential Loss of Life (PLL) method. This method will provide a statistical expectation of the annual number of casualties, which could be calculated in a conservative way for a specific area by multiplying the location-related risk with the population, or by means of a more realistic calculation based on the (partial) societal risk model for a specific area of influence, assuming protective factors.[9]

Standards regarding Location-Related Risk

The applied standard for a location-related risk amounts to housing and other vulnerable objects, such as schools and hospitals, in new situations (new risky activities or new land developments in the vicinity of existing activities) one in a million per year (ten-six per year).

The location-related risk is independent of the actual population in the vicinity of an airport. When determining the location-related risk, only fictitious persons on the ground are taken into consideration. The risk concerning occupants of an aircraft (the loss of life) is not part of the location-related risk.

Regarding existing situations, the use of a more flexible standard in the range of one in thousand per year will now be applied on housing and (other) vulnerable objects (ten-five per year). With respect to less vulnerable objects such as office buildings and business, a similar approach will be applied with a location-related risk standard, which is a tenfold higher.

Standards regarding Societal Risk

Regarding societal risk, no standards will be applied, although there is accountability. The accountability includes that the competent authority is required, besides the mathematical level of the societal risk, to take the following considerations along in its administrative balancing:

- The possibilities regarding self-rescuing ability of individuals in the vicinity of the risk source;
- The possibilities regarding the controllability of a serious incident or a disaster;
- Possible alternatives.
  and
- Possible risk reducing measures.

In addition, the following elements play an important role:

- Usefulness and necessity of the development/activity
  and
- Changes over time.

9 RIVM Report 620002001/2005, A.A.C. Van Vliet, E.S. Kooi & J.G. Post, Societal Risk and the Area-Specific Approach, 2 May 2005, at 8.

11.4 TOTAL RISK LOAD

The Total Risk Load (*Totaal risico gewicht TRG*) is a measure, which indicates the total external safety risk generated by air traffic at an airport. TRG is calculated over a period of one year and consists of the multiplication result of the number of aircraft movements, the average accident probability of these movements and the fleet size average maximum take-off mass of an aircraft.

Restricted Zone Schiphol

The Resolution Airport Allocation Schiphol (*Luchthavenindelingbesluit LIB* Schiphol 26 November 2002) includes, with respect to the external safety risk, legal standards concerning permitted building (construction) within the restricted zone around Schiphol international airport. These standards can be summarized as follows:

- The restricted zones within the location-related risk-contour larger than 10-5 may not be used for (existing) homes and other sensitive buildings.
- Within the restricted zones with a location-related risk of 10-6 it is not allowed to plan and build new buildings, unless:
- These buildings are Schiphol-related offices/businesses or logistical businesses.
- A declaration of no objection is issued.

At present, there are no regulations included in the resolution regarding the allocation and use of land within the airport perimeter, except the assignment of the runway structural system.

Other Airports in the Netherlands

Since the adoption of the legislation draft on Regulatory Civil Airports and Military Airports (*Regelgeving voor Burgerluchthavens en Militaire luchthavens RBML*) in November 2008, which presently is part of the Dutch Aviation Act, airport regulation has to be setup for all airports, including military airports (air force and maritime airbases).

In the airport regulation of civil airports, being of regional significance, rules have to be included regarding the allocation and use of land in connection with external safety risk. In the airport regulation of military airports, an ultimate limitation regarding the external safety risk can be included. The Dutch National Aerospace Laboratory developed a calculation methodology for military fixed-wing aircraft commissioned by the Minister of Defence, the so-called "Model parameters for the assessment of TPR around military airports".

11.5 Disasters and Major Accidents Act

The planning for disaster control at aerodromes in the Netherlands has its origins in the Disasters and Major Accidents Act (30 January 1985), the Resolution Disaster Control Aerodromes (16 July 2001) and the Fire Service Regulations on Civil Aerodromes (1974, amended in 2004, expired 20 January 2010).

The Disasters and Major Accidents Act requires that the College of mayor and aldermen be responsible for preparation regarding disasters and major accidents control within their municipality. In particular, the College should promote training and establish agreements necessary for an adequate control of disasters or major accidents. Based on Article 4 of this Act, the mayor establishes once every four years a disaster control plan with regard to disasters and major accidents whose location, nature and consequences are foreseeable.

After establishing a disaster-control plan, the mayor will submit this plan for review by the Provincial Governor. This dignitary will review the plan according to the criteria set out in Article 3 of the Resolution Disaster Control Airports as well as to the Resolution Quality Criteria regarding Planning of Disaster Control and the thereby-established review protocol. The Provincial Governor also has the statutory mandate to promote coordination in the planning phase, to establish the provincial coordination plan and to arrange the interregional assistance on the basis of the Fire Service Act of 30 January 1985.

The regional fire service has under the Fire Service Act of 1985 a statutory duty with respect to the preparation on coordinating assistance regarding disaster control and the preparation regarding the operational direction of an actual disaster.

The Resolution Disaster Control Aerodromes, based on the Disasters and Major Accidents Act (Articles 3 and 4), requires that the mayor, after consultation with the commander of the military airbase, establish a disaster-control plan for aircraft accidents at airports and the immediate surroundings (Article 2).

The mayor will send this plan to the Minister of Transport. In addition, Article 4 requires that the College of mayor and aldermen will provide for training exercises, with which accuracy, completeness and usability will be tested. With respect to the implementation of the evaluation, the military commander will be involved. Article 3 stipulates which requirements the contents of the disaster control plan must comply.

11.6 Special Fire Service Regulations on Civil and Military Aerodromes

The aerodrome operator is required to establish an emergency plan pursuant to Article 3.1 of the Fire Service Regulation on Civil Aerodromes 2004. An emergency plan has to comply with the provisions of this Regulation. This emergency plan consists of measures to be taken to be fully prepared in case of an aircraft accident or serious incident at the airport or in the immediate surrounding areas. The emergency plan includes a survey of procedures and methods to be used by the appropriate airport organization as well as external assistance and rescue organizations. This Regulation also includes provisions for training of this particular emergency plan and the evaluation of the training exercises. The Fire Service Regulation on Civil Aerodromes 2004 is based on international ICAO legislation concerning fire fighting at airports.[10]

Regarding military aerodromes, the requirements are laid down in the Dutch Air Traffic Regulations and various NATO-criteria laid down and maintained in Standard NATO agreements (STANAGs). The Royal Netherlands Air Force has a further elaboration in the Fire Service Regulation RNLAF (*Koninklijke Luchtmacht Klu*) VS-BRW and the Manual Flight Safety, Quality, Working Safety and Environment (VKAM). The Fire Service Regulation RNLAF VS-BRW defines roles and responsibilities of military fire department officers assigned to an air force section. Based on these rules, military aerodromes are required as well to establish an emergency plan. The requirements for such a plan will be broadly consistent with the relevant provisions of the Fire Service Regulation on Civil Aerodromes.

11.7 Concluding and Additional Remarks

The safety risk levels at airports are mainly dependent on the airside facilities such as runways, taxiways, aprons, navigational aids, radar, lighting and emergency response (firefighting equipment capability), and to a certain extent, the flight crew and air traffic control proficiency, the movement, the operational standard, the technical status and safety record of aircraft using the airports. The latter of these items, including conditions such as terrain and weather, have more effect on the risk levels in the vicinity of airports. In the past, a relatively low number of fatal accidents occurred in the vicinity of European airports. To mention the most striking:

On 27 November 1983, an Avianca Boeing 747-283B with registration HK-2910 crashed following a descent below the minimum descent altitude (MDA) approximately 12 km southeast of Madrid Barajas international airport. On 4 October 1989, an El-Al Boeing 747-258F Cargo aircraft crashed after an uncontrollable final flight segment following the separation of two engines in a populated area in Amsterdam (the so-called Bijlmer disaster). On 22 December 1999, a Korean Boeing 747-2B5F cargo aircraft flight 8509 (registration HL-7451) crashed shortly after take-off from London Stansted international airport close to but clear of local houses. The accident was caused by pilots' misinterpretation of malfunctioning, erratic flight instruments.

On 25 July 2000, an Air-France Concorde, flight 4590, crashed in the town of Gonesse seconds after its take-off from Paris Charles de Gaulle international airport, after a fuel

10 ICAO Annex 14 Chapter 9 Emergency and other services.

tank puncture by debris from a ruptured tire caused by a titanium alloy strip separated from a previously departed aircraft.

On 24 November 2001, an AVRO RJ100 (registration mark HB-IXM) of Crossair (flight LX-3597) crashed in hilly terrain only 4 km short of runway 28 of Zurich Kloten international airport. The accident investigation revealed controlled flight into terrain (CFIT) caused by a series of pilots errors and navigational faults that led the aircraft off-course, hitting trees on a hilltop just north of its assigned landing runway. On 25 February 2009, a Turkish Airlines Boeing 737-800, flight 1951, crashed during the landing procedure in a muddy field close to the approach lights of runway 18R of Schiphol international airport after a failed stall recovery procedure following malfunctioning aircraft instrument and system reactions (also known as the Poldercrash).

In order to limit exposure to human fatality or injury from aircraft crashes during approach, landing and take-off manoeuvres, development of public zones (location-related high risk areas) in the vicinity of airports, attracting many people for reasons unrelated to aviation, workers, commuters or residents, shall be restricted (TPR). Appropriate legislation based on calculated risk patterns, risk-assessment methodologies and statistical expectation of fatal accidents in the vicinity of airports and also economical, political and financial considerations are factors influencing the restrictive building and housing policy around busy major European airports.

12 Conclusions

Aviation safety oversight is the process of ensuring that airmen, air carriers, aircraft, manufacturers and numerous others who are engaged in aviation perform their functions in a safe and responsible manner. Safety oversight also ensures that the national aviation industry provides a safety level that is equal to or better than the requirements of the ICAO SARPs. The primary objective is to create the safest possible civil aviation system in order to acquire public trust and willingness to universally use air-transport system. However, to achieve an acceptable, or rather an idealistic high-quality-level of aviation safety throughout the world, a global regulatory regime for aviation activities is a prerequisite.

Following the Paris Convention of 1919, the Chicago Civil Aviation Conference created in 1944, during the Second World War, an international legal framework (rules and regulations) with respect to international air navigation, the *Convention on International Civil Aviation* (Chicago Convention).

At the same time, a permanent global international aviation organization ICAO, a specialized agency of the United Nations, was established to promote safe and orderly development of international civil aviation throughout the world. It sets standards and regulations necessary for aviation safety, security, efficiency and regularity, as well as for aviation environmental protection.

Annexes, being an integral part of the Chicago Convention and containing detailed aviation safety-related standards, recommended practices and associated procedures, were developed, extended, adopted and continually amended by ICAO.

Implementation of these internationally agreed standards is obligatory, due to the binding force, except for timely notification (compulsory under Article 38 Chicago Convention) of differences by contracting States unable to identify themselves with the multidisciplinary content of the requirements.

According to ICAO Annex 6, it is mandatory that contracting States have a system in place to monitor all flight-operation activities of its air carriers in international air transportation. In this respect, the State of the Operator has major obligations to ensure air-traffic safety in general, and more specifically the safe transport of passengers and cargo operated by the State's air carriers.

ICAO's adopted SARPs deals with practically all these activities. By implementing these regulations, the contracting States adhere to their obligations as signatory States to the Chicago Convention, and thus will establish a State's safety oversight system, which may be required to fulfil those obligations.

However, there are contrasting views regarding the legal obligation to implement or to comply with the provisions contained in the 18 Annexes, which form the basis for most of the world's national aviation regulations. On one hand, Annexes are considered to be (compulsory) regulations adopted by international agreement, but on the other hand, Annexes are circumstantial adaptable, expandable and amendable or flexible and thus open for practicable adoption.

The safe and orderly development of international civil aviation requires that all civil aviation operations, both on the ground and in the air, shall be conducted under internationally accepted minimum operating standards, procedures and practices. Implementation of SARPs by an ICAO contracting State must normally be effected under the rule of law promulgated by that State, which in turn will require an enactment of a specific legislative framework to be able to discharge its obligations and responsibilities and to develop and promulgate civil aviation regulations consistent with the acceptance by that State of the provisions of ICAO SARPs. This legislative framework also addresses the necessary powers to ensure compliance with these provisions.[1]

Two fundamental principles are crucial for aviation safety oversight: first of all, every State has complete and exclusive sovereignty over the airspace above and laterally bounded by its territory (Chicago Convention Article 1 Sovereignty and Article 2 Territory) and secondly, a State's responsibility for safety oversight is one of the tenets of the Chicago Convention and individual States should reaffirm their safety oversight responsibilities and obligations (ICAO Assembly Resolution A29-13). If these obligations are not or not fully respected by States, aviation-safety deficiencies may arise.

With respect to the European Union and its member States, together forming a new transnational structure based on the notion of mutual interdependence, the classical concept of indivisible sovereignty does not fit in the political and legal realities of the Community order, however not without inherent, deep-rooted resistance.

Within the European Union, a transfer of powers and responsibilities will take place. The member States determine which powers are transferred to the European Union, which undoubtedly will strengthen the functioning of the European Union, especially through the Treaty of Lisbon, signed on 13 December 2007, and which agreement established, among other things, the concept of shared competence in the transport sector (member States cannot exercise competence in areas where the European Union has done so).

In this perspective, the European Union has developed an air-transport policy from as early as 1977 until the present. A remarkable and unparalleled process because for the first time in the world civil aviation experience, a group of States together has developed a comprehensive mandatory air-transport policy including the adoption of rules and legislation, of which aviation safety is a vital element.

1 ICAO Doc 9734.

This EU legislation, essentially consisting of Regulations and Directives as legal instruments, will be applied with the force of national law and European law as appropriate, creating a unified safe air-transport market.

Harmonizing of these Regulations and Directives will ensure that common rules regarding licensing, airworthiness, operations of aircraft, accident and incident investigation, aeronautical products, parts and appliances, etc., are applicable throughout the European Union. Therefore, the objective of aviation-related Community legislation is primarily intended to enhance air-transport safety in all aspects.

As early as 1955, since the establishment of ECAC, this European cooperative aviation organization has the objective to implement internationally agreed safety standards in a uniform manner within the European region. The ECAC still plays an important role in developing and harmonizing civil aviation policy in Europe. ECAC collaborates closely with ICAO and the European Union, while contacts are maintained with regional civil aviation organizations and States outside Europe.

Parallel to the developments within ICAO regarding auditing safety oversight, ECAC examined the safety of foreign aircraft. ECAC introduced the voluntary SAFA programme to be able to build a European database of reports, incidents and findings, both technical and operational, of foreign aircraft using European airports.

The European Commission and the Parliament came to the same conclusion after some tragic aircraft accidents involving European citizens and expressed the need to take a more active stance and develop a strategy to improve the safety of its citizens travelling by air or living in the vicinity of airports.

The European Commission issued in 1996 a Communication to the European Parliament and the Council, SEC(96)1083, entitled "Defining a Community Aviation Safety Improvement Strategy", which clearly states that safety may be effectively enhanced by ensuring that aircraft fully comply with the international safety standards laid down within the Annexes to the Chicago Convention, with the overall intention to introduce a harmonized approach to the effective enforcement of these international safety standards within the Community in order to establish and maintain a high uniform level of civil aviation safety in Europe. Appropriate legislation has materialized this intention.

In 2001, the European Commission further outlined the strategy regarding the European Community contribution to world aviation safety improvement (COM(2001) 390 final). This strategy was based on two mainstreams of actions. On one hand, it was recommended to intensify work to maintain and further improve the safety of operations in Europe, including the creation of EASA. The second action plan was aiming at verifying that third countries actually do apply internationally agreed aviation-safety standards, and if applicable assisting them in complying. The result was the adoption of the so-called SAFA-Directive on the safety of foreign aircraft using Community airports so that appropriate information can be gathered and analyzed and actions taken by the Community and its member States to rectify dangerous situations caused by possible deficiencies. Following the adoption and coming into force of the SAFA-Directive, the former ECAC SAFA programme was transferred to the competence of European Community institutions, *inter alia* the European Commission and EASA as from 1 January 2007.

The European Commission was not the only entity or organization in the world confronted with this civil aviation-safety phenomenon. The U.S. FAA developed a similar foreign assessment programme (IASA), however, focusing on a country's ability, not the individual air carrier, to assess whether that State's civil aviation authority, overseeing its national air carriers that operate, or seek to operate routes to the United States, adheres to SARPs for aircraft operations and maintenance established by ICAO.

ICAO itself established USOAP in 1999 to carry out mandatory audits on contracting States to assist and advice these States to improve their capability of safety oversight by assessing whether the critical elements of a safety oversight system have been implemented effectively, so as to enable them to monitor their civil aviation sector in accordance with ICAO standards. In order to assist States in forming an opinion on the status of the safety oversight capability of audited States, ICAO made available transparent summary reports to all the contracting States, although not without previous opposition (on 16 July 2008 all audited States eventually consented the release of the results). These reports included an abstract of the audit findings, the corrective actions proposed by the State concerned, the actual status of implementation of SARPs, and comments by ICAO on the overall soundness of the safety oversight system in each audited State. USOAP is in a real sense still a voluntary programme because ICAO has no decisive legal power to oblige contracting States to accept the necessary and valuable inspections and audits regarding provisions of all safety-related Annexes (Comprehensive Systems Approach), while these States are committed to apply all the ICAO SARPs. It is an accomplished fact that ICAO can only opt for a formal bilateral Memorandum of Understanding with contracting States without creating any (international) right or obligation of a binding nature.

Notwithstanding the voluntary nature of the programme, and thus being entirely dependent on cooperative consent of the respective State, USOAP audits have demonstrated that accurate, transparent, factual and timely information will enable all contracting States and the aviation community in general, to develop a strategy to assist in the resolution of identified safety concerns. Contracting States have been unanimous in their recognition of the USOAP as an essential tool, which should continue to exist as one of the major activities of ICAO to contribute to the improvement of global civil aviation safety.

Improving civil aviation safety could be considered as the result of constant and collective contributions of individuals and organizations in different aviation disciplines. However, in order to achieve the best constructive results, tremendous efforts, time and planning are required because there is simply no easy way to instantaneously make aviation globally safe. It is rather a step-by-step process. One of the steps is to establish an outline for the practical global-safety umbrella that should ensure safety, regularity and efficiency of civil aviation operations.

National but above all international standardization is essential in all matters affecting operations of aircraft, airworthiness and the numerous facilities and services required in their support such as telecommunications, aerodromes, navigation aids, meteorology, air-traffic services, search and rescue, aeronautical information services and aeronautical charts. The next step is to achieve the highest practicable degree of uniformity worldwide whenever this will facilitate and improve civil aviation safety, efficiency and regularity based on safety-relevant ICAO SARPs, associated procedures, guidance material and Procedures for Air Navigation Services (PANS).

Developing and harmonizing aviation-safety oversight organizations and methods on State and international level should be accomplished as a positive contribution, and as a countermeasure, aviation-safety complacency should be avoided at all times. Further improvement of aviation safety will require constant investments by management in equipment, training, human resources, operational research and detailed analyzes of aviation accidents and incidents to be able to learn lessons for the future.

While aviation safety improvements will assume a more extensive form, the civil aviation world is still split in different categories: those States "being compliant with the provisions of the ICAO Annexes", and those States "not or not fully being compliant".

To protect the interest of the European citizens who may be living in the vicinity of major airports or travelling on board of foreign aircraft, the European Union identified the need to effectively enforce international safety standards within the Community, where most States already have the status of being compliant with the provisions of the Annexes to the Chicago Convention.

Enforcement is done through the conduct of inspections, including random ramp inspections on foreign aircraft landing at the airports located in EU member States. Enforcement is legally recognized on the basis of the provisions of the SAFA Directive.

The SAFA programme is a mandatory applied useful (bottom-up) instrument to gain insight into the safety condition and level of an aircraft, the air carrier and subsequently of a certain State's oversight capability. If this characteristic sequence is maintained, the SAFA programme could be considered as supplemental to more top-down oriented programmes, such as the ICAO USOAP and the FAA IASA. The SAFA programme comprises random ramp inspections or targeted inspections of aircraft of which EU member State authorities suspect that these third-country aircraft may not comply with ICAO standards. As a reminder, third-country aircraft are, in this context, considered as aircraft that are not used or operated under the control of a competent authority of a Community member State.

A major and legitimate feature of the SAFA programme is its non-discriminatory nature. However, both the programme and the subsequent EU operating ban mechanism, in particular designed to protect European air travellers and citizens on the ground, and more generally to enhance global aviation safety, also have triggered criticism. Safeguarding civil aviation safety through time-limited ramp inspections is assumed as an inexpedient method. Assessment of foreign aircraft in this manner will not reveal critical factors which only show up in flight conditions such as airmanship, pilot skills, situational awareness, distress handling etc. Other fundamental factors such as the safety culture of the respective air operator and the extent to which requirements are complied with, remain ostensibly hidden. Only pursuant to various sources of information, a solid assessment can be done.

There are concerns about the fact that serious political and economical interests are influencing the inclusion of air carriers in the EU blacklist because, for blacklisting, there are no defined global standards. In some cases, incumbent air carriers should gain advantage. International aviation bodies such as IATA assume that such a list will not contribute to solve the fundamental safety problems associated with certain foreign aircraft operations. They have criticized the blacklist for being punitive rather than constructive in terms of improving safety procedures.

The EU blacklist is dominated by Asian and African air carriers. Most of these carriers are not even flying to Europe, but the list is intended more generally to inform European travellers. Blacklisting is in the opinion of many disadvantaged stakeholders rather unfair, the reason why some of the banned air carriers find different ways (aircraft leasing) to operate to Europe. USOAP will be a more effective tool, as it has the potential to address core issues and global safety concerns.[2] The downside is that this programme requires the consent of the contracting State to be audited.

The European Commission is of the opinion that the application of Regulation (EC) No 2111/2005, in conjunction with the SAFA Directive, has already served as a clear demonstration of the value-added by the Community system in the field of aviation safety. Rather than being employed as a punitive instrument, the blacklist has proved to be an efficient dissuasive and pre-emptive measure, and it definitely spurs foreign air carriers to reform their operations.

The EU blacklist is capable of identifying *ex ante* air-safety deficiencies with potentially disastrous repercussions.[3] Inclusion in the blacklist is prompted by criteria such as an inappropriate or insufficient corrective action plan applied by an air carrier, or by its deplorable safety record, as well as the repeated lack of ability to timely address safety deficiencies because of continued query by EU member States. Ramp inspections and any subsequent operating bans are believed to be stringent measures, but provide irrefutable tangible results.

2 R.I.R. Abeyratne, *Strategic Issues in Air Transport, Legal, Economic and Technical Aspects*, Springer-Verlag Berlin and Heidelberg GmbH & Co. KG, New York 2012, at 156.

3 SEC (2009) 1735 final Commission Staff Working Document.

However, alleged disproportionate assessment and treatment by national aviation authorities regarding foreign air carriers using airports within the Community, resulting in suspension and revocation of air-carrier operations by some member States, prompted some of these carriers to file lawsuits, especially before 2006 when the European Union, through the careful assessment of the Aviation Safety Committee, adopted and published a Community-wide aviation blacklist of those air carriers banned from operating within the European Union. The list is based on the principle that air carriers banned in one member State are banned in all member States.

While the contracting States, signatories of the Chicago Convention, recognize that every State has complete and exclusive sovereignty over the airspace above its territory, these States are unable to individually regulate the safety of international civil aviation without the cooperation of all other States.

The European Union member States, although responsible for national aviation safety oversight, have developed a comprehensive mandatory Community air-transport policy.

This process is unique and completely different from the ICAO approach in the sense that particular importance has been attached to the economic aspects of such a policy, and in that the rules are mandatory without the possibility of member States filing differences. The European Union rules and regulations are there to be applied with the force of national law and European law as appropriate. Aviation safety is a vital element of this mandatory air-transport policy. The first priority of the SAFA programme and the Community blacklist is to maintain and enhance the already high level of aviation safety in Europe.

The EU blacklist and the SAFA programme have fostered a closer cooperation between the Community and non-EU States as well as international aviation-related organizations in order to improve international aviation safety oversight in general, and especially to verify compliance by air carriers with the relevant international aviation-safety standards.

There are however some restrictions: SAFA ramp inspections are only carried out within the Community, so far limiting the proposed global effort to improve civil aviation safety. Moreover, inclusion on the EU blacklist depends on available and verifiable information, while ailing and banned air carriers continue to fly to other regions in the world. In addition, in global or regional economic downturn, air carriers, being cyclical enterprises, have the tendency to cut back on aviation safety, the cornerstone of civil air transport. Fortunately, this civil air-transport system, being highly advanced and subject to strict supervision and international safety standards, has some safety redundancy, but there is a lower limit not to be exceeded in everyone's interest.

A fortiori the task of governments to strengthen their safety oversight capability, despite economic recession-related cutbacks in State expenditure, and the objective of aviation-related organizations to join forces in finding the right tools to enhance global aviation safety in the future.

Selected Bibliography

1 Treaties

Convention on International Civil Aviation, Chicago, 7 December 1944 (Chicago Convention) (ICAO Doc 7300/9).
Convention on the Regulation of Aerial Navigation, Paris, 13 October 1919.
Vienna Convention on the Law of Treaties, 23 May 1969.
Treaty of Lisbon Amending the Treaty on European Union (Maastricht Treaty) and the Treaty Establishing the European Community (Treaty of Rome), Lisbon, 13 December 2007.

2 Cases

Case related to Phuket Airlines Co. Ltd., Thailand.
LJN BA 2577 Rechtbank 's-Gravenhage, case number: AWB 05/8479, 21 March 2007.
Case related to Onur Air Tasimacilik A.S., Turkey.
LJN BB 6588 Rechtbank 's-Gravenhage, case number: AWB 05/6303, 29 October 2007.
Case related to Phuket Airlines Co. Ltd., Thailand.
LJN BC 3585 Raad van State, case number: 200703161/1, 6 February 2008.

3 Documents

1. European Union Documents

Council Regulation (EEC) No 3922/91 of 16 December 1991 on the harmonization of technical requirements and administrative procedures in the field of civil aviation, *Official Journal of the European Communities L* 373, 31 December 1991.
Council Regulation (EEC) No 2407/92 of 23 July 1992 on licensing of air carriers, *Official Journal of the European Communities L* 240/1, 24 August 1992.
Council Directive No 94/56/EC of 21 November 1994 establishing the fundamental principles governing the investigation of civil aviation accidents and incidents, *Official Journal of the European Union L* 319/14, 12 December 1994.
SEC (96) 1083 final, 12 June 1996. Defining a Community aviation safety improvement strategy. Report by the High Level Group established by the Council Decision of 11th March 1996 addressed to the European Parliament and to the Council. Communication from the Commission to the Council and the European parliament.

SEC (96) 1083-C4-0423/96, 20 March 1997. Report on the Commission's communication on defining a Community aviation safety improvement strategy.

Regulation (EC) No 1049/2001 of the European Parliament and of the Council of 30 May 2001 regarding public access to European Parliament, Council and Commission documents, *Official Journal of the European Communities L* 145/43, 31 May 2001.

Regulation (EC) No 1592/2002 of the European Parliament and of the Council of 15 July 2002 on common rules in the field of civil aviation and establishing a European Aviation Safety Agency, *Official Journal of the European Union L* 240/1, 7 September 2002.

Directive 2003/42/EC of the European Parliament and of the Council of 13 June 2003 on occurrence reporting in civil aviation, *Official Journal of the European Union L* 167/23, 4 July 2003.

Regulation (EC) No 1643/2003 of the European Parliament and the Council of 22 July 2003 amending Regulation (EC) No 1592/2002 on common rules in the field of civil aviation and establishing a European Aviation Safety Agency, *Official Journal of the European Union L* 245/7, 29 September 2003.

Commission Regulation (EC) No 1702/2003 of 24 September 2003 laying down implementing rules for the airworthiness and environmental certification of aircraft and related products, parts and appliances, as well as for the certification of design and production organizations, *Official Journal of the European Union L* 243/6, 27 September 2003.

Commission Regulation (EC) No 2042/2003 of 20 November 2003 on the continuing airworthiness of aircraft and aeronautical products, parts and appliances, and on the approval of organizations and personnel involved in these tasks, *Official Journal of the European Union L* 315/1, 28 November 2003.

Commission Regulation (EC) No 1701/2003 of 24 September 2003 adapting Article 6 of Regulation (EC) No 1592/2002 of the European Parliament and of the Council on common rules in the field of civil aviation and establishing a European Aviation Safety Organization, *Official Journal of the European Union L* 243/5, 27 September 2003.

Commission Regulation (EC) No 104/2004 of 22 January 2004 laying down rules on the organization and composition of the Board of Appeal of the European Aviation Safety Agency, *Official Journal of the European Union L* 16/20, 23 January 2004.

Directive 2004/36/EC of the European Parliament and of the Council of 21 April 2004 on the safety of third-country aircraft using Community airports, *Official Journal of the European Union L* 143/76, 30 April 2004.

Regulation (EC) No 2111/2005 of the European Parliament and of the Council of 14 December 2005 on the establishment of a Community list of air carriers subject to an operating ban within the Community and on informing air transport passengers of the identity of the operating air carrier, and repealing Article 9 of Directive 2004/36/EC, *Official Journal of the European Union L* 344/15, 27 December 2005.

Commission Regulation (EC) No 473/2006 of 22 March 2006 laying down implementing rules for the Community list of air carriers which are subject to an operating ban within the Community referred to in Chapter II of Regulation 2111/2005 of the European Parliament and of the Council, *Official Journal of the European Union L* 84/8, 23 March 2006.

Commission Regulation (EC) No 474/2006 of 22 March 2006 establishing the Community list of air carriers which are subject to an operating ban within the Community referred to in Chapter II of Regulation (EC) No 2111/2005 of the European Parliament and of the Council, *Official Journal of the European Union L* 84/14, 23 March 2006.

Commission Regulation (EC) No 707/2006 of 8 May 2006 amending Regulation (EC) No 2042/2003 as regards approvals of a limited duration and Annexes I and III, *Official Journal of the European Union L* 122/17, 9 May 2006.

Commission Regulation (EC) No 768/2006 of 19 May 2006 implementing Directive 2004/36/EC of the European Parliament and of the Council as regards the collection and exchange of information on the safety of aircraft using Community airports and the management of the information system, *Official Journal of the European Union L* 134/16, 20 May 2006.

Commission Regulation (EC) No 334/2007 of 28 March 2007 amending Regulation (EC) No 1592/2002 of the European Parliament and the Council on common rules in the field of civil aviation and establishing a European Aviation Safety Agency, *Official Journal of the European Union L* 88/39, 29 March 2007.

Commission Regulation (EC) No 376/2007 of 30 March 2007 amending Regulation (EC) 2042/2003 on the continuing airworthiness of aircraft and aeronautical products, parts and appliances, and on the approval of organizations and personnel involved in these tasks, *Official Journal of the European Union L* 94/18, 4 April 2007.

Regulation (EC) No 216/2008 of the European Parliament and of the Council of 20 February 2008 on common rules in the field of civil aviation and establishing a European Aviation Safety Agency, and repealing Council Directive 91/670/EEC, Regulation (EC) No 1592/2002 and Directive 2004/36/EC, *Official Journal of the European Union L* 79/1, 19 March 2008.

Commission Regulation (EC) No 351/2008 of 16 April 2008 implementing Directive 2004/36/EC of the European Parliament and of the Council as regards the prioritisation of ramp inspections on aircraft using Community airports, *Official Journal of the European Union L* 109/7, 19 April 2008.

Commission Directive 2008/49/EC of 16 April 2008 amending Annex II to Directive 2004/36/EC of the European Parliament and of the Council regarding the criteria for the conduct of ramp inspections on aircraft using Community airports, *Official Journal of the European Union L* 109/17, 19 April 2008.

Commission Regulation (EC) No 1056/2008 of 27 October 2008 amending Regulation (EC) No 2042/2003 on the continuing airworthiness of aircraft and aeronautical products, parts and appliances, and on the approval of organizations and personnel involved in these tasks, *Official Journal of the European Union L* 283/5, 28 October 2008.

Commission Regulation (EC) No 690/2009 of 30 July 2009 amending Regulation (EC) No 216/2008 of the European Parliament and of the Council on common rules in the field of civil aviation and establishing a European Aviation Safety Agency, and repealing Council Directive 91/670/EEC, Regulation (EC) No 1592/2002 and Directive 2004/36/EC, *Official Journal of the European Union L* 199/6, 31 July 2009.

Regulation (EC) No 1108/2009 of the European Parliament and of the Council of 21 October 2009 amending Regulation (EC) No 216/2008 in the field of aerodromes, air traffic management and air navigation services and repealing Directive 2006/23/EC, *Official Journal of the European Union L* 309/51, 24 November 2009.

Commission Regulation (EC) No 1194/2009 of 30 November 2009 amending Regulation (EC) No 1702/2003 laying down implementing rules for the airworthiness and environmental certification of aircraft and related products, parts and appliances as well as for certification of design and production organizations, *Official Journal of the European Union L* 321/5, 8 December 2009.

SEC (2009) 1735 final, 11 January 2010. Commission Staff Working Document. On the report from the Commission to the Council and the European Parliament regarding the application of Regulation (EC) No 2111/2005.

Commission Regulation (EU) No 127/2010 of 5 February 2010 amending Regulation (EC) No 2042/2003 on the continuing airworthiness of aircraft and aeronautical products, parts and appliances, and on the approval of organizations and personnel involved in these tasks, *Official Journal of the European Union L* 40/4, 13 February 2010.

Regulation (EU) No 996/2010 of the European Parliament and of the Council of 20 October 2010 on the investigation and prevention of accidents and incidents in civil aviation and repealing Directive 94/56/EC, *Official Journal of the European Union L* 295/35, 12 November 2010.

Commission Regulation (EC) No 1071/2010 of 22 November 2010 amending Regulation (EC) No 474/2006 establishing the Community list of air carriers which are subject to an operating ban within the Community, *Official Journal of the European Union L* 306/44, 23 November 2010.

Commission Regulation (EU) No 1178/2011 of 3 November 2011 laying down technical requirements and administrative procedures to civil aviation aircrew pursuant to Regulation (EC) No 216/2008 of the European Parliament and of the Council, *Official Journal of the European Union L* 311/1, 25 November 2011.

Commission Implementation Regulation (EU) No 1197/2011 of 21 November 2011 amending Regulation (EC) No 474/2006 establishing the Community list of air carriers which are subject to an operating ban within the Community, *Official Journal of the European Union L* 303/14, 22 November 2011.

COM (2000) 595 final, Proposal for a Regulation of the European Parliament and of the Council on establishing common rules in the field of civil aviation and creating a European Aviation Safety Agency, 27 September 2000.

COM (2001) 390 final, Communication from the Commission. A European Community contribution to World Aviation Safety Improvement, 16 July 2001.

COM (2002) 8 final, Proposal for a Directive of the European Parliament and of the Council on the safety of third countries aircraft using Community airports, 14 January 2002.

COM (2002) 649 final, Communication from the Commission on the consequences of the Court judgements of 5 November 2002 for European air transport policy, 19 November 2002.

COM (2005) 579 final, Proposal for a Regulation of the European Parliament and of the Council amending Regulation (EC) No 1592/2002 of 15 July 2002 on common rules in the field of civil aviation and establishing a European Aviation Safety Agency, 15 November 2005.

COM (2009) 627 final, Report from the Commission to the European Parliament and the Council, European Union SAFA Programme, 18 November 2009.

COM (2009) 710 final, Report from the Commission to the Council and the European Parliament on the application of Regulation (EC) No 2111/2005 regarding the establishment of a Community list of air carriers subject to an operating ban within the Community and informing air transport passengers of the identity of the operating air carrier, and repealing Article 9 of Directive 2004/36/EC, 11 January 2010.

COM (2011) 159 final, Report from the Commission to the European Parliament and the Council, European Union SAFA Programme, 31 March 2011.

EASA Decision No 2008/001/S of the Executive Director of the European Aviation Safety Agency of 29 September 2008 on the Guidance Material for the Qualification of SAFA Inspectors.

2. *ICAO Documents*

ICAO Doc 8335 AN/879 *Manual of Procedures for Operations Inspection, Certification and Continued Surveillance.*

ICAO Doc 9318 *Protocol Relating to an Amendment to the Convention on International Civil Aviation (Article 83 bis).*

ICAO Doc 9284 AN/905 *Technical Instructions for the Safe Transport of Dangerous Goods by Air,* 2011-2012 edn.

ICAO Doc 9376 *Preparation of an Operations Manual,* 2nd edn, 2000.

ICAO Doc 9587 *Policy and Guidance Material on the Economic Regulation of International Air Transport,* 3rd edn, 2008.

ICAO Doc 9734 AN/959 *Safety Oversight Manual, Part A, The Establishment and Management of a State's Safety Oversight System*, 2nd edn, 2006.

ICAO Doc 9735 AN/960 *Safety Oversight Manual,* 2nd edn, 2006.

ICAO Doc 9760 *Airworthiness Manual Volume I & II*, 1st edn, 2005.

ICAO Assembly Resolution A29-13 Improvement of Safety Oversight 1992.

ICAO Assembly Resolution A32-11 Establishment of an ICAO Universal Safety Oversight Audit Programme, 1 October 1998.

ICAO Assembly Resolution A35-6 Transition to a comprehensive systems approach for audits in the ICAO Universal Safety Oversight Audit Programme (USOAP), 6 October 2004.

ICAO NACC/DCA/2-WP/07, 18 August 2005, Second meeting of North American, Central American and Caribbean Directors of Civil Aviation, Tegucigalpa, Honduras, 11-14 October 2005, International Aviation Safety Data Exchange (IASDEX) Programme.

ICAO NACC/DCA/2-WP/04, 31 August 2005, Second meeting of North American, Central American and Caribbean Directors of Civil Aviation, Tegucigalpa, Honduras, 11-14 October 2005, Progress Report on the Implementation of the ICAO USOAP.

ICAO Working Paper High-Level Safety Conference, HLSC 2010-WP/78, Montreal 29 March-1 April 2010, Doc 9935.

ICAO Assembly Working Paper A23-WP/16 EX/4 Amendment to the Chicago Convention with respect to problems arising out of lease, charter and interchange of aircraft in international operations (proposed Article 83 bis), 21 April 1980.

ICAO Assembly Working Paper A33-WP/8 EC/3 Report of the Council on Financial and Organizational Aspects of Airports and Air Navigation Services, 11 June 2001.

ICAO Assembly Working Paper A33-WP/9 EC/4 Report by the Council on Joint Ventures in the Collection of Air Navigation Services Charges, 11 June 2001.

ICAO Assembly Working Paper A33-WP/15 EX/2 Improving SARPs by Focusing on System-Level and Performance Requirements, 14 June 2001.

ICAO Assembly Working Paper A33-WP/47 EX/10 Report on the ICAO Universal Safety Oversight Audit Programme, 9 July 2001.

ICAO Assembly Working Paper A33-WP/48 EX/11 Continuation and Expansion of the ICAO Universal Safety, 9 July 2001.

ICAO Assembly Working Paper A33-WP/49 EX/7 Assisting States to Resolve Deficiencies Identified by the Safety Oversight Audits, 9 July 2001.

ICAO Assembly Working Paper A33-WP/50 EX/14 Report of the Council on Implementation of Assembly Resolution on Increasing the Effectiveness of ICAO, 25 July 2001.

ICAO Assembly Working Paper A33-WP/61 EX/18 A European View of the Value of the ICAO USOAP and Its Expansion, 24 July 2001.

ICAO Assembly Working Paper A33-WP/62 EX/19 ECAC AVSEC Airport Audit Programme, 24 July 2001.

ICAO Assembly Working Paper A33-WP/66 EX/23 Enhancing the Quality and Usefulness of ICAO Audit Information, 25 July 2001.

ICAO Assembly Working Paper A33-WP/83 EX/33 IATA Operational Safety Audit (IOSA), 9 August 2001.

ICAO Assembly Working Paper A33-WP/112 EX/39 Need for the Development of the Safety Oversight Concept and a Long-Term Work Programme in This Field, 29 August 2001.

ICAO Assembly Working Paper A35-WP/152 EX/64 TE/20 Airport Safety Oversight, 23 September 2004.

ICAO Assembly Working Paper A35-WP/167 EX/67 Airport Security, 23 September 2004.

ICAO Assembly Working Paper A36-WP/25 Activities of the International Financial Facility for Aviation Safety (IFFAS), 5 July 2007.

ICAO Assembly Working Paper A36-WP/64 EX/22 Progress Report on the Implementation of the ICAO USOAP under the Comprehensive Systems Approach, 2 August 2007.

ICAO Assembly Working Paper A37-WP/273 TE/146 Coordinating the Activity of Participating Countries Implementing Safety Assessment Programmes for Foreign Aircraft and Using the Results of these Inspections, 27 September 2010.

ICAO Working Paper APIRG/15-WP/9 Report on the Implementation of the Comprehensive Systems Approach for the Conduct of Safety Oversight Audits under ICAO USOAP, 27 July 2005.

ICAO Working Paper DGCA/06-WP/8 Directors General of Civil Aviation Conference on a Global Strategy for Aviation Safety, Montreal 20-22 March 2006. Mutual Recognition, 9 January 2006.

ICAO Working Paper DGCA/06-WP/3 Directors General of Civil Aviation Conference on a Global Strategy for Aviation Safety, Montreal 20-22 March 2006. The Status of Safety Oversight, 9 January 2006.

3. *Other Documents*

AAIASB Air Accident Investigation & Aviation Safety Board, *Air Accident Report Helios Airways Flight HCY 522 Boeing 737-31S at Grammatiko, Hellas on 14 August 2005.*

ANSV Agenza Nazionale per la Sicurezza del Volo, *Final Report Accident Involving ATR 72 Aircraft Registration Marks TS-LBB, Ditching Off the Coast of Capo Gallo (Palermo-Sicily) August 6th, 2005.*

Committee on Aircraft Certification Safety Management, Aeronautical and Space Engineering Board, Commission on Engineering and Technical Systems, National Research Council, *Improving the Continued Airworthiness of Civil Aircraft. A Strategy for the FAA's Aircraft Certification Service*, National Academy Press, Washington, DC 1998.

Dutch Expert Group Aviation Safety (DEGAS) Advisory report 2011-034, *Veiligheid als deel van het geheel – een rationele bedding voor emotie*, 4 July 2011.
Enquêtecommissie vliegramp Bijlmermeer, *Een beladen vlucht, final report of the Bijlmer Enquête,* Sdu Publishers, The Hague 1999.
FAA Federal Aviation Administration 14 CFR (Code of Federal Regulation) Part 129 Certification Process Document for Foreign Air Carriers.
FAA 14 CFR Part 91 General Operating and Flight Rules.
JIAAG Junta Investigadora de Accidentes de Aviación Civil, *Informe Final West Caribbean Airways DC-9-82 (MD-82) Matrícula HK 4374X Machiques, Venezuela, 16 de Agosto de 2005.*
NTSB National Transportation Safety Board Aircraft Accident Report PB91-910404 NTSB/AAR-91/04, *Avianca, the Airline of Columbia, Boeing 707-321B, HK 2016 Fuel Exhaustion, Cove Neck, New York, January 25, 1990.*
NTSB Aircraft Accident Report PB97-910406 NTSB/AAR-97/06 DCA96MA054, *In-Flight Fire and Impact with Terrain ValuJet Airlines Flight 592 DC-9-32, N904VJ, Everglades, near Miami, Florida, May 11, 1996.*
OECD Organisation for Economic Co-operation and Environment, *Globalization, Transport and the Environment,* 2010.
Republic of South Africa, *Report of the Board of Inquiry into the Loss of South African Airlines Boeing 747-244B Combi Aircraft into the Indian Ocean on November 28th 1987.*
RIVM Report 62000 2001/2005, A.A.C. Van Vliet, E.S. Kooi & J.G. Post, Societal Risk and the Area Specific Approach (*Groepsrisico en gebiedsgerichte benadering*), 2 May 2005.
The Kingdom of Saudi Arabia, General Authority of Civil Aviation, Safety and Economic Regulations, Safety Department, *Aircraft Accident Report MD-11F King Khalid International Airport Riyadh Abnormal Runway Contact (ARC) 15 Sha'aban 1431 H – 27 July 2010 G.*
University of Bielefeld-Faculty of Technology, *Bericht der Dirrectión General de Aeronáutica Civil (DGAC) der Dominikanischen Republik über die untersuchung des unfalles mit dem flugzeug B-757 am 06, Februar 1996 bei Puerto Plata.*

4 *Related Books and Articles*

R.I.R. Abeyratne, *Strategic Issues in Air Transport, Legal, Economic and Technical Aspects,* Springer-Verlag Berlin and Heidelberg GmbH & Co. KG, New York 2012.
L.H.D. Bakker, 'Risky Carriers or Risky States?', *in* H.M. Soekkha (Ed.), *Aviation Safety: Human Factors – System Engineering – Flight Operations – Economics – Strategies – Management*, VSP BV, Zeist 1997.
L.J. Brinkhorst, *Europese Unie en nationale soevereiniteit* (oration Leiden University), 2008.

A.J. Broderick & J. Loos, 'Government Aviation Safety Oversight – Trust, but Verify', 67 *JALC* 2002, p. 1035.

R.W. Cobb & D.M. Primo, *The Plane Truth, Airline Crashes, the Media, and Transportation Policy*, Brookings Institution, Washington, DC 2003.

B. Crumley, 'Does the E.U.'s Airline Blacklist Make Flying Safer?', *TimeWorld*, 2009.

E. Denza, 'From Aerostats to DC-10s-Recognition of Certificates of Airworthiness', *in* A. Kean (Ed.), *Essays in Air Law*, Martinus Nijhoff Publishers, The Hague 1982.

F. de Florio, *Airworthiness: An Introduction to Aircraft Certification*, Butterworth-Heinemann (Elsevier), Oxford 2006.

J. Huang, *Aviation Safety through the Rule of Law, ICAO's Mechanisms and Practices*, Kluwer Law International, Alphen aan den Rijn 2009.

J. Huang, *Aviation Safety and ICAO* (Ph. D. dissertation, Leiden University, Leiden), Kluwer Law International, Alphen aan den Rijn 2009.

A. Knorr, *Will 'Blacklists' Enhance Airline Safety?*, German Aviation Research Society Workshop, Amsterdam 2006.

A. Kotaite, 'Universal Audit Programme Conducted under Auspices of ICAO Would Offer Many Benefits', *ICAO Journal*, Vol. 53, No. 4, 1998, pp. 5-6.

J. Learmont, 'Global Airline Review of 2009', *Flight International*, 2010.

J.H. Loos, The ICAO Safety Oversight Program, The George Washington Aviation Institute, 1999.

S. Michaelides-Mateou & A. Mateou, *Flying in the Face of Criminalization. The Safety Implications of Prosecuting Aviation Professionals for Accidents*, Ashgate Publishing Limited, Farnham 2010.

M. Milde, 'Enforcement of Aviation Safety Standards – Problems of Safety Oversight', 45 *Zeitschrift für Luft- und Weltraumrecht* 1996, pp. 3-17.

M.A. Piers, 'Models for the Assessment of Third Party Risk due to Aircraft Accidents in the Vicinity of Airports', *in* A. Mosleh & R.A. Bari (Eds.), *Probabilistic Safety Assessment and Management*, Vol. III, Springer-Verlag London Ltd., London 1998.

A.L.C. Roelen, *Causal Risk Models of Air Transport, Comparison of User Needs and Model Capabilities*, IOS Press, Amsterdam 2008.

W. Rosenkrans, *Delving into IOSA*, AeroSafetyWorld Flight Safety Foundation, December 2010-January 2011.

B. Scherer, 'Der Fall Onur Air', *Frankfurter Allgemeine Zeitung*, 19 May 2005.

H.J. Schlamp, S. Simons, H. Stark & G. Traufetter, 'Flying Coffins. How Strict Is the EU's Airline Blacklist?', *Der Spiegel*, 2009.

A.J. Stolzer, C.D. Halford & J.J. Goglia, *Safety Management Systems in Aviation*, Ashgate Publishing Company, Burlington, VT 2008.

B. Straus, *Europe Lifts Ban on Phuket Airlines, Adds PIA to Updated Blacklist*, Air Transport World, 2007.

U. Uchim, *Lack of Cooperation Killing African Carriers – Chingosho, Secretary-General/ CEO, AFRAA*, The Sun Publishing, 2011.

A. Walker, 'Inclusion on EU Aviation Blacklist Spurs Reform in Central Asia', EU-Central Asia Monitoring Commentary No 9, 2010.

J.M. Walters & R.L. Sumwalt III, *Aircraft Accident Analysis: Final Reports*, McGraw-Hill Professional, New York, NY 2000.

L. Weber, 'Convention on International Civil Aviation-60 Years', 53 *Zeitschrift für Luft- und Weltraumrecht* 2004, p. 289.

A.T. Wells & C.C. Rodrigues, *Commercial Aviation Safety*, 4th edn, McGraw-Hill Professional, New York, NY 2004.

L. Werfelman, *Making a List*, AeroSafety World Flight Safety Foundation, 2009.

L. Werfelman, *Speak Up*, AeroSafety World Flight Safety Foundation, 2010.

About the Authors

Dick van het Kaar (1948) has worked in the aviation industry for over 44 years. He is a former Royal Netherlands Air Force helicopter and jet fighter pilot officer and instructor. In 1977 he joined the world of civil aviation and has logged more than 23,000 flying hours. During his flying career as a senior captain B747 at KLM/Martinair he studied International Law at the Open University in the Netherlands, where he earned his degree of Master of Laws.

Dick van het Kaar was also member of the Appeal Council of the Dutch Airline Pilots Association and is author of several Aviation Law-related essays, books and articles in various international Air and Space Law journals. Since July 2009 he has been associated with historic aircraft restoration projects at the Dutch National Air and Space Museum.

Dr. Ronald Schnitker (1950) has worked in the field of aviation law enforcement for over 30 years. He served as Commanding Police Officer at the Aviation Investigation Department of the Dutch National Police Force, and later as Director Legal Affairs at Eindhoven Airport, and is currently working as an independent lawyer.

In April 2011 he was appointed Vice-President of the Royal Netherlands Aeronautical Association. He has published widely, particularly on Enforcement of Aviation Law.

Ronald Schnitker is chief editor of *Journaal Luchtrecht*, a Dutch Aviation Law magazine. He has been assigned by the Netherlands Civil Aviation Authority as an expert on theoretical Aviation Law examinations. He obtained a juris doctor degree from the University of Tilburg.